Ali Al-Abadi

Novel Strategies for Performance Improvement of Wind Turbines

Ali Al-Abadi

Novel Strategies for Performance Improvement of Wind Turbines

Südwestdeutscher Verlag für Hochschulschriften

Impressum / Imprint
Bibliografische Information der Deutschen Nationalbibliothek: Die Deutsche Nationalbibliothek verzeichnet diese Publikation in der Deutschen Nationalbibliografie; detaillierte bibliografische Daten sind im Internet über http://dnb.d-nb.de abrufbar.
Alle in diesem Buch genannten Marken und Produktnamen unterliegen warenzeichen-, marken- oder patentrechtlichem Schutz bzw. sind Warenzeichen oder eingetragene Warenzeichen der jeweiligen Inhaber. Die Wiedergabe von Marken, Produktnamen, Gebrauchsnamen, Handelsnamen, Warenbezeichnungen u.s.w. in diesem Werk berechtigt auch ohne besondere Kennzeichnung nicht zu der Annahme, dass solche Namen im Sinne der Warenzeichen- und Markenschutzgesetzgebung als frei zu betrachten wären und daher von jedermann benutzt werden dürften.

Bibliographic information published by the Deutsche Nationalbibliothek: The Deutsche Nationalbibliothek lists this publication in the Deutsche Nationalbibliografie; detailed bibliographic data are available in the Internet at http://dnb.d-nb.de.
Any brand names and product names mentioned in this book are subject to trademark, brand or patent protection and are trademarks or registered trademarks of their respective holders. The use of brand names, product names, common names, trade names, product descriptions etc. even without a particular marking in this work is in no way to be construed to mean that such names may be regarded as unrestricted in respect of trademark and brand protection legislation and could thus be used by anyone.

Coverbild / Cover image: www.ingimage.com

Verlag / Publisher:
Südwestdeutscher Verlag für Hochschulschriften
ist ein Imprint der / is a trademark of
OmniScriptum GmbH & Co. KG
Heinrich-Böcking-Str. 6-8, 66121 Saarbrücken, Deutschland / Germany
Email: info@svh-verlag.de

Herstellung: siehe letzte Seite /
Printed at: see last page
ISBN: 978-3-8381-5107-6

Zugl. / Approved by: FAU, Diss., 2014

Copyright © 2015 OmniScriptum GmbH & Co. KG
Alle Rechte vorbehalten. / All rights reserved. Saarbrücken 2015

Abstract

In this thesis, the influence of the turbulence on the performance of the Horizontal Axis Wind Turbine (HAWT) has been investigated. For that numerical optimizations for aerodynamic shape design, pitch-control, analysis and semi-empirical performance predictions are developed. These methods are numerically and experimentally validated.

First, a turbine Torque-Matched Aerodynamic Shape Optimization method (TMASO) which maximizes the power while matching the drive unit torque has been developed. This method helps in producing an efficient laboratory-scale turbine that is used later in the experimental phase of the present work for the turbulence investigations. Second developed method is the Torque-Matched Pitch Control Optimization (TMPCO) which is aimed at keeping the rated power constant by adjusting the pitch angle. It changes the wind turbine's control strategy from a stall- to a pitch-control one. In addition, an analysis method for Torque-Matched Aerodynamic Performance Analysis Method TMAPAM has been developed to predict the turbine performance.

The reliability of the design and the analysis methods were proven by applying them with additional modifications to large scale wind turbines. They are used for improving and predicting the performance of different HAWTs scales and control strategies. The stall-regulated NREL 10kW and the pitch-controlled NREL 5MW wind turbines are used as baselines for these processes. The validation of the stall-regulated turbine needs beside the consideration of the turbine properties a post-stall model to predict the lift and drag coefficients. In the present study, modifications of existence post-stall models and experimental observations have been done to produce a semi-empirical post-stall model. It can be applied to predict the aerodynamic characteristics with considering the HAWT scale and design.

A special designed setup that can measure the performance in the wind tunnel has been developed. The generator torque needed for the calculation is measured separately with a second experimental setup. The comparison of the experimental results with that of the analysis method showed a good agreement.

Additional validations of the developed methods were conducted via RANS numerical simulations with STAR-CCM+. Furthermore, the flow over the wind turbine blade at different operating points are assessed in regards of separation and three dimensional flow on the blade. This provides a deep understanding of the flow behaviour over the blades.

Finally, experimental investigations have been conducted by exposing the developed efficient wind turbine model to different turbulence levels. The turbulence is generated by using two static squared grids (Fine and Coarse). The developments of turbulence scales in the flow direction at various Reynolds numbers and the grid mesh size are measured.

Detailed measurements and analysis of the upstream and downstream turbulence intensity (TI) distributions and the spectrum analysis are conducted to give insight on the surrounding free stream and turbine wake interaction and how can different turbulence levels influence the performance of the turbine. Performance measurements are conducted with and without turbulence grids and the results are compared.

The study has shown the higher the turbulence level, the higher the power coefficient. There are a few mechanisms responsible for this behaviour. First, is the interaction of turbulence scales with the blade surface boundary layer, which in turn delay the stall. Thus, suppressing the boundary layer and preventing it from separation and hence enhancing the aerodynamics characteristics of the blade. In addition, higher turbulence helps in damping the tip vortices. Thus, reduces the tip losses. Adding winglets to the blade tip will reduce the tip vortex, and then it is possible to isolate major part of them.

Furthermore, high turbulence content serves in increasing the wake-surrounding interaction, and hence more energy entrainment to the wake region. More energetic turbulent flow has been shown to penetrate through the turbine blades, which brings more power in the near-wake region that cause a faster recovering of wake.

Zusammenfassung

In dieser Arbeit wurde der Einfluss der Turbulenz auf die Leistung einer Horizontalachswindturbine (HAWT) untersucht. Hierzu wurden numerische Optimierungen für das aerodynamische Schaufeldesign, Neigungswinkelsteuerung, Analyse und semi-empirische Leistungsvorhersage entwickelt. Diese Verfahren wurden numerisch und experimentell validiert.

Zunchst wurde eine Optimierungsmethode für die Turbinenschaufelform (TMASO) entwickelt, welche die Leistung unter Berücksichtigung des Drehmoments der Generatoreinheit maximiert. Diese Methode ermöglicht das Erstellen einer effizienten Windturbine im Labormaßstab, die im experimentellen Teil dieser Arbeit fr Untersuchungen zur Turbulenz Verwendung findet. Die zweite entwickelte Methode ist die TMPCO Optimierung, die darauf abzielt, die Nennleistung der Turbine durch Anpassung des Neigungswinkels konstant zu halten. Diese verndert die Steuerungsstrategie fr die Turbine von einer Strömungsabrisssteuerung zu einer Neigungswinkelsteuerung. Fr die Vorhersage der Turbinenleistung wurde ein Nachrechnungsverfahren (TMAPAM) entwickelt.

Die Zuverlässigkeit der Entwicklungs- und Nachrechnungsverfahren wurde durch deren Anwendung auf Windturbinen im großen Maßstab gezeigt. Sie werden benutzt, um die Leistung von Horizontalachswindturbinen mit unterschiedlichem Maßstab und Steuerungsstrategie vorherzusagen und zu verbessern. Die strmungsabrissgeregelte 10kW-Turbine und die neigungswinkelgesteuerte 5MW NREL Phase-VI Windturbine werden als Basis fr diese Prozesse genutzt. Die Validierung der strmungsabrissgeregelten Turbine benötigt neben der Berücksichtigung der Turbineneigenschaften ein Modell fr die Vorhersage der Auftriebs- und Widerstandsbeiwerte nach dem Strmungsabriss. In der vorliegenden Arbeit wurden hierzu Modifikationen von vorhandenen Modellen und experimentelle Beobachtungen verwendet, um ein modifiziertes, semi-empirisches Modell nach dem Strömungsabriss zu entwickeln. Es kann nicht nur zur Vorhersage der aerodynamischen Eigenschaften verwendet werden, sondern auch, um den

Maßstab der Windturbine und deren Design zu berücksichtigen.

Ein speziell entworfener Messaufbau, mit dem die Leistung in einem Windkanal gemessen werden kann, wurde entwickelt. Das Drehmoment des Generators, das für die Berechnung bentigt wird, wird separat mit einem zweiten experimentellen Aufbau ermittelt. Der Vergleich der Versuchsergebnisse mit denen der Analysemethode zeigte eine gute bereinstimmung. Zusätzliche Validierungen der entwickelten Methoden wurden mittel numerischer Simulationen (RANS) mit STAR-CCM+ durchgeführt. Des Weiteren wurde die Umstrmung der Windturbinenschaufel bei unterschiedlichen Betriebspunkten hinsichtlich Strömungsablösung und der dreidimensionalen Strömung auf der Schaufel ausgewertet. Dies liefert ein grundlegendes Verständnis des Strömungsverhaltens um die Schaufel bei verschiedenen Betriebspunkten. Zuletzt wurden experimentelle Untersuchungen am entwickelten, effizienten Windturbinenmodell bei unterschiedlichen Turbulenzgraden durchgeführt.

Die Turbulenz wurde hierbei durch zwei statische Rechteckgitter (fein und grob) erzeugt und die Entwicklung der Turbulenzskalen in Strömungsrichtung bei verschiedenen Reynoldszahlen und Gittergrößen gemessen. Detaillierte Messungen und Analysen der Verteilungen der Turbulenzintensität (TI) stromauf- und stromabwärts sowie eine Spektralanalyse wurden durchgeführt, um Einblick in die Interaktion zwischen der freien Außenströmung und des Turbinennachlaufs zu geben und wie unterschiedliche Turbulenzgrade die Leistung der Turbine beeinflussen knnen. Leistungsmessungen der Turbine mit und ohne Turbulenzgitter wurden durchgefhrt und die Ergebnisse verglichen.

Die Untersuchung zeigte, dass der Leistungsbeiwert der Turbine umso höher ist, je höher der Turbulenzgrad ist. Zunchst ist es hier die Interaktion der Turbulenzskalen mit der Grenzschicht auf der Schaufeloberflche, welche den Strömungsabriss verzögert. Demnach bewirkt eine Unterdrckung der Grenzschicht und das Verhindern ihrer Ablsung eine Verbesserung der aerodynamischen Eigenschaften der Schaufel. Außerdem wirkt sich ein höherer Turbulenzgrad durch die Dmpfung des Blattspitzenwirbels positiv aus, was wiederum die Verluste verringert. Das Anbringen von Winglets an der Schaufelspitze wrde den Blattspitzenwirbel reduzieren und es somit mglich machen, einen Großteil davon für weitere Untersuchungen zu unterdrcken. Jedoch ist es nicht möglich, durch Winglets den Spitzenwirbel komplett zu unterdrcken. Darüber hinaus unterstützt ein hoher Turbulenzgrad die Interaktion zwischen Turbinennachlauf und Umgebung und führt damit zu einem höheren Energieeintrag in das Nachlaufgebiet. Es hat sich gezeigt, dass mehr energiereiche turbulente Strömung den Rotor durchdringt, was mehr Energie in das rotornahe Nachlaufgebiet bringt und somit zu einer schnelleren Zurückbildung des Nachlaufs führt.

Acknowledgements

I would like to express my gratefulness to the German Academic Exchange Service (DAAD) and the Iraqi MoHESR who offered me the Fellowship at 2010.

It is an honour for me to present my thanks mainly to Prof. Dr. Antonio Delgado for his acceptance of supervising my thesis, which was the first spark in this work. He has made his support in a number of ways during my work. I appreciate providing the required research facilities in his institute. In addition, the direct and friendly communications with his students has made this work smoothly finish. This thesis would not have been possible without the great support offered by my co-supervisor Asst. Prof. Dr. Özgür Ertunc. He has been always beside, adding his invaluable suggestions, monitoring the stations of my work and sharing with me his wide knowledge and experience.

I appreciate the valuable advices and discussions offered by Prof. Dr. Philipp Epple. I am particularly grateful for the technical consultant and assistance provided by Mr. Horst Weber in the electrical workshop of the institute.

I would like to thank the valuable discussions with Prof. Dr. Jovan Jovanovic and Mr. Hermann Lienhart.

Special thanks are extended to my colleagues in the group and in the institute: Haresh Vaidya, Manuel Muensch, Jens Krauss, Cagatay Köksoy, Oliver Litfin and others. They were always ready for help.

It is a pleasure to thank those who have helped me in the administration section; Rita Scheffler-Kohler, Sonja Hupfer, and in the IT section; Thorsten Beilke and Sebastian Röhl. I owe my steepest gratitude to my parents who have always been standing beside. To my father Prof. Khedher Al-Abadi, who has inspired and oriented the major part of my life and my mother Sabiha who has spiritually supported me. I am grateful to my brothers Dr. Firas and Ahmed.

This thesis is dedicated to my wife Abeer and my daughters Rafal and Ronza who have always stood by me.

Contents

1. **Introduction and Outline**   1
   1.1. General on HAWT's 1
   1.2. Background on Numerical Simulation 3
   1.3. Background on Turbulence 4
   1.4. Objectives 6
   1.5. Methodology and Thesis Structure 7

2. **Theoretical Background**   10
   2.1. Wind Power and Aerodynamics 10
      2.1.1. Betz limit 10
   2.2. Airfoil Theory 13
      2.2.1. Forces on airfoils 13
      2.2.2. Drag coefficient components 15
      2.2.3. Separation and Stall 15
   2.3. Design Theories 16
      2.3.1. Design theory of Betz 16
      2.3.2. Design theory of Schmitz 19
      2.3.3. Blade Element Momentum (BEM) theory 21
      2.3.4. Lifting-Surface Prescribed-Wake Performance Prediction Method 23
      2.3.5. Efficiency of the wind turbine 24
   2.4. Wind Turbines Control Strategies 25
   2.5. Optimization 28
      2.5.1. Evolutionary Methods 28
      2.5.2. Gradient-Based Optimization 29
      2.5.3. Evolutionary versus Gradient 32
   2.6. Characterization of Turbulence 33
      2.6.1. Turbulence Intensity 33
      2.6.2. The Auto-Correlation Function 33
      2.6.3. Scales of Turbulent Flows 34

	2.6.4. Spectrum of Turbulence	36

3. Numerical Set-up — 37
3.1. Governing Equations 37
3.2. RANS Equations 39
3.3. Setting up the Problem 40
    3.3.1. Geometry 40
    3.3.2. Grid generation 41
    3.3.3. Physical Model 43
        3.3.3.1. Near Wall Treatment 44
        3.3.3.2. Moving Reference Frame 45
        3.3.3.3. Overset Mesh 45
    3.3.4. Specification of Boundary Conditions 46

4. Experimental Facilities — 47
4.1. Wind tunnel 47
4.2. Turbulence Generating Grids 48
    4.2.1. Fine Grid 48
    4.2.2. Coarse Grid 49
4.3. Setup for Performance Measurement 50
4.4. Setup for Turbulence measurement 52
    4.4.1. Instrumentations 52
    4.4.2. Investigation Setup 53
    4.4.3. Calibration of the Hot-Wire 54

5. Optimization Procedure — 56
5.1. Optimization of a laboratory scale WT 56
    5.1.1. Aerodynamic Design procedure 57
    5.1.2. Formulation of the shape optimization 59
    5.1.3. Formulation of the pitch-control optimization 63
5.2. Optimization of NREL stall-regulated WT 66
    5.2.1. Settings of Input Parameters 66
    5.2.2. Shape optimization of 10kW stall-regulated HAWT 68
    5.2.3. Pitch control optimization of stall-regulated WT 69

6. Performance Analysis — 71
6.1. Performance Analysis of Laboratory Scale WT 71
6.2. Performance analysis of the stall-regulated WT 76
6.3. Modified Post-stall model 84

6.4. Performance prediction of the stall-regulated shape optimized WT	87
6.5. Performance Analysis of the Pitch-Control Optimized HAWT	88
6.6. NREL 5MW Wind Turbine	91
6.6.1. Performance analysis of NREL 5MW	91

7. Numerical Validations — 96

7.1. Power Coefficient	96
7.1.1. Tangential Velocity on 2-D Blade Sections	97
7.1.2. Pressure Coefficient Distribution on Sections	99
7.1.3. 3-D Flow on Blade	101
7.2. VS-VP analysis	102
7.2.1. VS-VP HAWT Performance	102
7.2.2. 3-D Flow on Blade of VS-VP	103
7.2.3. Pressure Coefficient Distribution for VS-VP	104

8. Experimental Investigations on the Influence of Turbulence — 106

8.1. Grid-Generated Turbulence	106
8.2. Measurements of the Power Coefficient	112
8.2.1. Performance Measurements With Fine Grid	112
8.2.2. Performance Measurements with Coarse Grid	113
8.3. Velocity Distribution	114
8.3.1. Velocity Distribution at Free Flow Turbulence (Without Grid)	114
8.3.2. Velocity Distribution at Medium Turbulence Level (With Fine Grid)	118
8.3.3. Velocity Distribution at High Turbulence Level (With Coarse Grid)	120
8.4. Influence of Turbulence on Tip Vortex	122
8.4.1. Tip Analysis	123
8.5. The Effect of Winglets	126
8.6. Turbulence Measurements	129
8.7. Spectrum Analysis	131

9. Conclusions and Outlook — 134

9.1. Conclusions	134
9.2. Outlook	137

A. Appendix — 147

A.1. Specifications of the NREL UAE phase-VI	147

B. Appendix — 151

B.1. Specifications of NREL 5MW	151

List of Figures

2.1. Distribution of wind velocity and pressure over rotor plane. 12
2.2. Airfoil aerodynamic characterizations 14
2.3. Angles and velocities on the airfoil before and at the rotor plane (Betz triangles). 17
2.4. Rotor sections. 18
2.5. Velocities triangles before and at the rotor plane. 19
2.6. Locus of torques of $C_{P,max}$, $C_{P,Betz}$ and of rated power over angular speed. 26
2.7. Different control strategies 26
2.8. Different control strategies 27
2.9. Illustration of the fluctuating velocity [14] 33
2.10. Illustration of the auto-correlation function [14]. 34

3.1. Geometrical configuration of inner domain and overset domain 41
3.2. Geometrical configuration of outer domain 41
3.3. Mesh distribution 42

4.1. Wind Tunnel of LSTM-Erlangen 48
4.2. Schematic of the experimental rig in the wind tunnel 48
4.3. The fine grid installed at the wind tunnel outlet 49
4.4. The coarse grid installed at the wind tunnel outlet 49
4.5. Wind turbine in the wind tunnel 50
4.6. Power measuring circuit 51
4.7. Setup for measuring mechanical power 52
4.8. Schematic functionality diagram of a CTA 54
4.9. Calibration curves of the hot-wire 55

5.1. Distribution of a and a' with the rotor radius ratio 59
5.2. Aerodynamic characteristics at low Reynold's number of 65000 and $\alpha = 8^o$ 61
5.3. TMASO optimized wind turbine 62
5.4. Torque-Matched aerodynamics shape optimization procedure diagram . 63

List of Figures

5.5. Insertion of β_c to reduce α and thus tangential force and torque during increasing wind velocity. 64
5.6. Pitch control angle distribution of TMPCO over wind speed for the laboratory scale wind turbine. 65
5.7. TMPCO procedure diagram. 66
5.8. Power coefficient over angle of attack calculated by TMASO. 67
5.9. Glide ratio over angle of attack calculated by TMASO. 67
5.10. Initial and optimized chord length and twist angle distributions along the blade by TMASO. 69
5.11. Pitch control angle distribution of TMPCO of $10kW$ and NREL $5MW$ wind turbine [55]. 70

6.1. Analysis flow diagram 73
6.2. Comparison between the calculated and the measured power coefficients 74
6.3. Comparison between rotor and drive torques versus the rotational speed 75
6.4. Torque matched velocity and rotational speed relation 75
6.5. Angle of attack (α) distribution along the rotor radius ratio for different tip speed ratios 76
6.6. Experimental lift coefficient of the S809 airfoil. A: "attached flow regime"; B: "high lift, stall development regime, dynamic stall"; C: "flat plate, fully stalled regime" [61]. 79
6.7. Experimental drag coefficient of the S809 airfoil [61]. 79
6.8. Performance validation procedure diagram of the stall-regulated turbine. 80
6.9. Calculated torque in comparison with other prediction methods of the NREL UAE phase-VI rotor[61]. 82
6.10. Calculated axial induction factor and experimental data of the aero-elastic code PHATAS of the NREL UAE phase-VI rotor [80]. 82
6.11. Calculated angle of attack and experimental data of the aero-elastic code PHATAS of the NREL UAE phase-VI rotor [80]. 82
6.12. Calculated angle of attack and prediction method of LSWT and measuring method of LFA of the NREL UAE phase-VI rotor [80]. 82
6.13. Comparison between the calculated and the experimental power coefficient over wind velocity of the NREL UAE phase-VI rotor [61]. 83
6.14. Comparison between the calculated and the experimental power coefficient over tip speed ratio of the NREL UAE phase-VI rotor [61]. 83
6.15. Calculated relative angle over blade length of the NREL UAE phase-VI rotor. 84

List of Figures

6.16. Calculated angle of attack over blade length of the NREL UAE phase-VI rotor. 84
6.17. Calculated axial induction factor over blade length and wind velocity of the NREL UAE phase-VI rotor. 84
6.18. Calculated tangential induction factor over blade length and wind velocity of the NREL UAE phase-VI rotor. 84
6.19. Model and equation torque for different number of blades (B=1,2,3). . . 86
6.20. Model and equation torque for different chord ratios ($c/c_i = 1, 1.25, 1.5, 1.75$). 86
6.21. Model and equation torque for different turbine radius ($R = 5, 10, 15, 20, 40m$). 86
6.22. Mean angle of attack versus wind velocity for different radius. 87
6.23. Effect of radius on the torque over wind velocities. 87
6.24. Power increment of the shape optimized turbine as compared to the baseline turbine experimental data [61]. 88
6.25. Power coefficient increment of the shape optimized turbine as compared to the baseline turbine experimental data [61]. 88
6.26. Calculated angle of attack over blade length of the pitch optimized NREL UAE phase-VI rotor. 89
6.27. Pitch-control performance analysis diagram. 90
6.28. Comparison between the calculated and the experimental power of the pitch optimized NREL UAE phase-VI rotor [61]. 90
6.29. Comparison between the calculated and the experimental power coefficient over tip speed ratio of the pitch optimized NREL UAE phase-VI rotor [61]. 90
6.30. Airfoil profiles and sections of the NREL $5MW$ wind turbine from $r = 12m$ to the tip. 92
6.31. Comparison between the calculated and the experimental power of the NREL $5MW$ wind turbine [55]. 93
6.32. Calculated angle of attack over blade length of the NREL $5MW$ wind turbine. 94
6.33. Calculated phi over blade length of the NREL $5MW$ wind turbine. . . . 94
6.34. Calculated lift coefficient over blade length and wind velocity of the NREL $5MW$ wind turbine. 94
6.35. Calculated drag coefficient over blade length and wind velocity of the NREL $5MW$ wind turbine. 94
6.36. Calculated axial induction factor over blade length and wind velocity of the NREL $5MW$ wind turbine. 94
6.37. Calculated tangential induction factor over blade length and wind velocity of the NREL $5MW$ wind turbine. 94

List of Figures

6.38. Comparison between the calculated and the experimental power coefficient over wind velocity of the NREL 6MW wind turbine [60]. 95
6.39. Comparison between the calculated and the experimental power coefficient over tip speed ratio of the NREL 6MW wind turbine[55, 60]. 95

7.1. Power coefficient over tip speed ratio 97
7.2. Tangential velocity on 2-D blade sections with velocity streamlines at low, design and high angle of attack 98
7.3. Residual for turbulent kinetic energy over tip speed ratio λ 99
7.4. Pressure coefficient on blade sections 100
7.5. Pressure distribution and blade streamlines at different angles of attack . 101
7.6. VS-VP control strategy 103
7.7. Pressure distribution and blade streamlines for VS-VP at wind velocities $8-16m/s$ 104
7.8. Pressure coefficient on blade sections for VS-VP 105

8.1. Intensity of grid-generated turbulence along the test section for the three cases (no grid, fine grid, coarse grid) 107
8.2. Distribution of Spectra $E(f)$ with eddies frequencies f at different oncoming wind velocities v_{wt} and turbulence levels for changing hot-wire positions x 108
8.3. Length scale distribution at medium turbulence level (fine grid) 109
8.4. Length scale distribution at high turbulence level (coarse grid) 109
8.5. Ratio λ/L at medium turbulence level (fine grid) 110
8.6. Ratio λ/L at high turbulence level (coarse grid) 110
8.7. Distribution of Re_λ at medium turbulence level (fine grid) 110
8.8. Distribution of Re_λ at high turbulence level (coarse grid) 111
8.9. Trend of C_P for the $S809$ profile at varying turbine axial positions with fine grid 113
8.10. Trend of C_P for the $SG6043$ profile at varying turbine axial positions with fine grid 113
8.11. Trend of C_P for the $S809$ profile at varying turbine axial positions with coarse grid 114
8.12. Trend of C_P for the $SG6043$ profile at varying turbine axial positions with coarse grid 114
8.13. Upwind velocity distribution without grid 115
8.14. Downwind velocity distribution without grid 116
8.15. Wake's outer and inner symmetrical borders without grid 116

List of Figures

8.16. Upwind velocity distribution along the axis of rotation ($y/D = 0$) without grid 117
8.17. Typical power coefficient curve and the operating points at different wind velocities 117
8.18. Upwind velocity distribution with fine grid 119
8.19. Downwind velocity distribution with fine grid 119
8.20. Wake's outer and inner symmetrical borders with fine grid 119
8.21. Upwind velocity distribution along the axis of rotation with fine grid . . 120
8.22. Upwind velocity distribution with coarse grid 121
8.23. Downwind velocity distribution with coarse grid 121
8.24. Wake's outer and inner symmetrical borders with coarse grid 121
8.25. Upwind velocity distribution along the axis of rotation with coarse grid 122
8.26. Tip vortices at $x/D = 0.2$ for different turbulence levels 123
8.27. Tip vortices at $x/D = 0.4$ for different turbulence levels 124
8.28. Tip vortices at $x/D = 0.6$ for different turbulence levels 125
8.29. Comparison of the tip vortex for a single revolution of the rotor at varying x/D and turbulence level 126
8.30. Winglet design employed in the experiments. Dimensions are in mm . . 127
8.31. The effect of winglet on C_P at free flow turbulence (without grid). All cases at wind turbine position of x=120cm 128
8.32. Downwind velocity distribution without grid and with winglets 129
8.33. Wake's outer and inner symmetrical borders without grid and with winglets 129
8.34. Downwind turbulent intensity distribution at different oncoming turbulence levels for the wind turbine position of 100 cm 130
8.35. Distribution of Spectra $E(f)$ with eddies frequencies f at different radial distance y/D in the turbine wake for different oncoming upwind turbulence levels at hot-wire downwind position of $x/D = 0.3$ 132
8.36. Distribution of Spectra $E(f)$ with eddies frequencies f at different radial distance y/D in the turbine wake for different oncoming upwind turbulence levels at hot-wire downwind position of $x/D = 1.1$ 133

A.1. S809 Airfoil profile [4]. 148
A.2. Chord and twist distributions over blade length 148
A.3. UAE phase-VI turbine mounted in NASA Ames $24.4m$ x $36.6m$ wind tunnel [72]. 148
A.4. UAE phase-VI rotor blade and pressure tabs defined as dotted lines [63]. 149
A.5. Power coefficient over tip speed ratio for the UAE phase-VI rotor at different tip pitch angles [34]. 149

List of Figures

A.6. Experimental torque and power over wind velocity of the UAE phase-VI rotor at a tip pitch angle of $3°$ [61] 150
B.1. 5MW wind turbine at Hooksiel off the German North Sea coast [7]. . . . 152
B.2. Chord distribution of the NREL $5MW$ wind turbine over blade length. . 153
B.3. Twist distribution of the NREL $5MW$ wind turbine over blade length. . 153
B.4. Airfoils profiles of the NREL $5MW$ wind turbine from $r = 11.75m$ to the tip [60]. 154
B.5. Pitch control angle distribution of the NREL 5MW wind turbine [55]. . . 154

List of Tables

3.1. Mesh cell and face count 43
3.2. Boundary conditions for wake and winglet model 46
3.3. Boundary conditions for pitch model 46

5.1. Comparison between initial and optimized values by TMASO. 69

6.1. Reynolds number distribution along the blade at $72rpm$ rotational speed and some wind velocities for UAE phase-VI turbine [63] 77

B.1. Distributed blade aerodynamic properties of the NREL $5MW$ wind turbine [55]. 153

Abbreviations

BEM	Blade Element Momentum
DOWEC	Dutch Offshore Wind Energy Converter
LFA	Local Flow Angle
LSWT	Lifting-Surface Prescribed-Wake
NREL	National Renewable Energy Laboratory
PHATAS	Program for Horizontal Axis wind Turbine Analysis and Simulation
TMASO	Torque Matched Aerodynamic Shape Optimization
TMPCO	Torque Matched Pitch Control Optimization
TMAPAM	Torque Matched Aerodynamic Performance Analysis Method
UAE	Unsteady Aerodynamxics Experiment
WindPACT	Wind Partnerships for Advanced Component Technology

Symbols

AR	aspect ratio
B	number of blades
C_D	drag coefficient
C_{Di}	induced drag coefficient
C_{D0}	zero-drag coefficient
C_F	axial thrust coefficient
C_L	lift coefficient
$C_{L,d}$	lift coefficient at design angle of attack
C_n	normal force coefficient
C_P	power coefficient
C_T	torque coefficient
C_t	tangential force coefficient
F	Prandtl's tip loss correction factor
F_D	drag force
F_L	lift force
F_y	tangential force

F_x	axial force
GR	glide ratio
Re	Reynolds number
P	power of the turbine
R	radius of the blade
T	torque of the turbine
T_N	rated torque of the turbine
dP	power on blade element
dT	thrust on blade element decomposed on the rotor axis
dU	torque on blade element decomposed on the rotor plane
a	axial induction factor
a'	tangential induction factor
c	chord length
e	blade span efficiency
n	rotational speed
p_0	ambient pressure
p_l	local pressure
r	radius at spanwise locations
u	tip speed
v_1	wind velocity long in front of the rotor
$v_{1,o}$	optimum wind velocity
v_N	rated wind velocity
v	wind velocity at the rotor plane
v_l	local wind velocity
v_3	wind velocity far behind the rotor
w	relative wind velocity
α	angle of attack
α_d	design angle of attack
β_0	tip pitch angle
β_t	twist angle
β	summation of tip pitch angle and twist angle
β_c	pitch control angle

Γ	bound circulation
γ	angle of relative wind to rotor axis
η	efficiency
λ	tip speed ratio
λ_o	optimum tip speed ratio
ρ	air density
φ	angle of relative wind to rotor plane
ω	angular speed
ω_o	optimum angular speed

1
Introduction and Outline

1.1. General on HAWT's

Horizontal axis wind turbines (HAWT's) are designed to extract the maximum power from the wind under many constraints arising from the employed technology and the environmental conditions. Designing of the HAWT involves a relatively complex geometry and subjected to multiple operating conditions. In different design steps of the HAWT there are different optimization objectives, such as the higher aerodynamic efficiency, lighter structures, lower fatigue loads, noise and cost [40]. These objectives are normally subjected to different constraints and trade-off among them to obtain the optimum one has to be considered. The efficiency of the wind turbines depends on many subsystems such as rotor shape, gear box, electrical generator and control of the turbine. The aerodynamics of the rotor blades shape playing a decisive role in maximizing the efficiency. Thus, Increasing energy demand requires more and more optimized rotors for efficient exploitation of the available wind energy.

Jureczko [57], devised a gradient based aerodynamic and structural shape optimization for the wind turbine blades. The optimization problem was subjected to constraints of the blade chord length, the tip speed and the optimal blade weight. Evolutionary optimization algorithms with embedded aerodynamic simulators were also utilized for the multi-objective blade shape optimization [32, 66, 17, 26].

1. Introduction and Outline

There are different types of control Strategies for commercial wind turbines. It can be stall regulated-fixed pitch (SR-FP) which is coupled to the generator via gearbox or direct-drive. The second control approach is to control the magnitude of the rotor current by adjustable external load. For example, Muljadi et al. [70], evaluated the variable speed fixed-pitch control (VS-FP) strategy to maximize the efficiency under varying wind speed conditions. The third approach is the active pitch control which can provide a wide range of rotational speeds [96]. Pitch-control strategy was also integrated in a multi-objective rotor shape optimization procedures [22].

In general, there are only two main major systems for controlling a wind turbine. These are blade pitch control, which changes the orientation of the blades to regulate the aerodynamic forces, and the generator torque control that is done by means of power electronics converter. Each control has its own objectives, such as power regulation, which in turns aim to obtain as much energy out of wind, or it could be a speed regulation that aims to restrict the noise by limiting the tip speeds of wind turbines. The coupling technology of the rotor with the generator has to be considered in the design. The use of the direct-drive turbines increases in comparison to the traditional gear-box ones as the wind industry needs scaling up, cutting costs, and improving reliability [9], and reducing the complexity owing to easier operations and maintenance [1, 8]. For any of the aforementioned drives and control technologies, the best performance of the wind turbine can be achieved when the aerodynamic shape of the turbine is matched to the torque rotational speed characteristic of the driving unit. This matching is more essential for the direct-drive technology, since the rotor is strictly constrained to the torque rotational speed characteristic of the generator.

Normally, wind turbines are designed for a specific wind condition at a specific site, which is so-called the operation point. However, turbines operate with reduced efficiency over a range of wind conditions different from the assumed operation point. Hence, the performance of the designed turbines has to be tested over range of possible wind conditions. The turbine performance analysis methods can be conducted experimentally, analytically, numerically or a hybrid of all. Many experimental measurement methods were established for different rotor diameters to verify the wind turbine performance characteristics [51, 52]. In general, nacelle-mounted anemometry are used to estimate the wind speed and incident. Therefore, the measurements involve challenges such as the site calibration under varying atmospheric conditions, instrument accuracy and corrections to air parameters. Even the performance measurement of small scale turbines with a few meters of rotor diameter can be very hard and costly as they have to be tested in wind-tunnels, which should have large enough cross sectional area or through truck tests [10].

The one-dimensional Blade Elemet Momentum (BEM) method [46] divides the blade into spanwise elements and ignores the interaction between them. It calculates the forces on each blade element by using the integral transport equations for mass, momentum and angular momentum in combination with the airfoil theory. The BEM method was proven to provide realistic results for steady conditions [92]. Based on the BEM theory a mathematical model for the fluid dynamics design of a wind turbine with correct evaluation of the induction factors has been implemented for evaluation of wind turbine rotor performance [61]. In contrast, the panel method gives insight into the three- dimensional wake flow and it can be more realistic in performance calculations with the inclusion of boundary layer approximations. The ongoing aerodynamic research is trying to address all the aspects that affect the operation performance such as three dimensionality, wake-surrounding air interaction, separation, stall, etc.. De Vries [33] provide a good overview of analytical and some numerical performance analysis methods based on the propeller theory and panel methods, respectively.

For the unsteady three-dimensional flow, the effect of the induced velocity around the HAWT rotor as well as the unsteady interaction of the rotor with upwind wake in array configuration can be well analyzed by the solution of Reynolds-Averaged Navier-Stokes (RANS) equations [23, 41, 38, 82]. Nevertheless, they have their own challenges such as the effort for grid generation, resolving the near wall region, selecting the appropriate turbulence model and sometimes demanding computational resources in terms of time and hardware. However, in many cases the numerical analysis methods are better choice for performance analysis, since complete similarity of large turbines can not be obtained in laboratory conditions.

1.2. Background on Numerical Simulation

Computational Fluid Dynamics (CFD) methods are widely applied in wind turbines investigations. They offer the possibility to study the flow over large structures, without the need for scaling. Further more, they are effective and practical when using in providing insights to flow phenomena such as three-dimensional, unsteady prediction, stall, boundary layer, flow separation and surface flow analysis. However, the quality and effectiveness of the computations are dependent on the applied physical models. Sumner [87], submitted a detailed review of the development of CFD applications in the wind energy community for various scales.

1. Introduction and Outline

CFD methods involve many turbulent models for different flow applications. Carcangiu [29] performed steady state Reynolds Averaged Navier-Stokes (RANS) computations with the $k-\omega$ Shear Stress Transport (SST) turbulence model to study the rotational effects on the boundary layer behaviour of the wind turbine blades. Mahu [65] modelled and simulated the NREL Phase IV rotor with the same model. Anjuri [21] applied this turbulence model with an added transition model by Langtry and Menter for simulations of the NREL Phase IV rotor, with good agreement to the published data. Laursen [62] modelled an Siemens $SWT-2.3-93$ variable speed wind turbine with the RANS $k-\omega$ SST model and showed a good torque match compared to field measurements. Kirrkamm [59] applied this turbulence model but performed the simulations as Unsteady Reynolds Averaged Navier-Stokes (URANS) in order to get a more realistic insight into the separated flow regions on the blade. Although the number of mesh cells was chosen low ($3.5*10^6$), the computational effort was considered elaborate. Carcangiu [28] additionally carried out wake investigations with RANS simulations. Due to the turbulent character of wakes, especially in the far wake, Vermeer [90] recommend unsteady Large Eddy Simulations (LES) or Detached Eddy Simulations (DES) for investigations. Wussow et al.[93] applied a LES to study the wake of the *Enercon E*66 turbine. They applied a turbulent wind field inlet and compared it to corresponding field measurements over a time period of several minutes. Simulations were matched the measurements quite well.

Winglets applications in wind turbines attracted only few researchers. For instance, Johansen [54] and Gaunaa [43] were performed RANS simulations to study the effect of winglets on wind turbines rotors and showed the potential to enhance the aerodynamic efficiency. In general, from the aforementioned, the numerical methods are robust when the flow is steady and uniform, otherwise, when using of the LES or the DNS, which in turn more expensive in terms of effort, time and cost. Thus, for complicated flows that involve extreme separation, flow interaction, stall and boundary layer detachment or turbulence, the experimental methods are preferred.

1.3. Background on Turbulence

It is common to design a wind turbine by using of the aerodynamics tools with considering of a uniform incoming flow without existence of turbulence. However, wind turbines normally operates under the influence of the turbulent atmospheric boundary layer. Thus, wind turbines are exposed to different levels of turbulence depending on the site and the time. Therefore, understanding the impact of the turbulence levels on the performance of the wind turbine as well as the downstream wake are crucial for the efficient design.

The oncoming wind to HAWTs change its speed and direction stochastically in time. Hence, turbine blades are exposed to flows with both fluctuating angle of attack and fluctuating yaw angles. The modern wind turbines are reacting to those changes by pitch angle and torque control not only to exploit as much power as possible but also to stabilize energy production and prevent any damage of the turbine. However, time scales of wind fluctuations and sudden changes of wind properties can be very short and very high in amplitude.

The effect of atmospheric and wake turbulences have been investigated in many studies. Barthelmie [74] proved that for a large offshore wind farm the losses in the power due to turbine wakes can reach up to 20%. Hansen [49] showed that at a specified wind farm the ambient turbulent intensity's has a linear influence on increasing the total wind farm efficiency for the wind speed range 7-8 m/s. Further study in a wind farm was conducted by Türk [89], which proved both power output and loads on turbine blades are increasing with the increment of turbulence intensity levels.

Many studies have shown that turbulence helps in improving the wake recovery due to enhanced mixing. Furthermore, turbulence serves in delaying the flow stall on the suction surface of the blade. Swalwell et al. [12], studied the influence of turbulence intensity on the aerodynamics performance of the thick airfoil NACA0021, which is normally placed at the root of wind turbine blades that is the first stalled part of the blade, by subjecting it to different levels of turbulence intensities. As a result, the increasing turbulence was found to delay stall until higher angle of attack in a way that is consistent with the delayed stall seen on HAWT. This was confirmed by Delnero [56] for a different airfoils and by Watkins [77] for low Reynols number wings due to the impact of the turbulence intensity on the laminar separation bubbles. Kamada [94] was experimentally proved that the turbulence flow prevents the flow separation on the airfoil surface and hence delays the stall. Whereas in another experimental work conducted by Maeda [18], it is found that the main flow turbulence has a direct impact on the wind turbine wake due to entrainment, which helps to recover the velocity deficit. Therefore, higher turbulence intensity leads to higher power output of the downstream wind turbine.

In addition to the incoming wind turbulence, the effect of the upstream turbine on the wake velocity and added turbulence intensity is still obvious even at a distance of

1. Introduction and Outline

fifteen times the rotor diameter, as presented by Chamorro [31]. The added wake turbulence has been estimated in many semi-empirical models. Frandsen [39] proposed a model for estimating the added turbulence by taking into account the wind-farm layout and additional surface roughness generated by the turbines. Whereas, Wessel [16] developed a semi-empirical model for calculating turbulence intensity inside offshore wind farms, which is taken from the wind speed deficit profile. It takes into account the wake superposition, interference and meandering to calculate the incidence of the turbulence intensity on each rotor within the farm for various wind direction and speed.

Additional challenges associated with turbulence such as unsteady aerodynamics forces, were investigated by Luhur [64]. His study aimed to model the dynamic lift and drag stochastically under turbulent conditions. This analysis brings further insight into the high frequency lift and drag dynamics. Changes in the angle of attack at time scales of one second and less was investigated [86]. The main focus laid on high frequent fluctuations in the wind and the resulting changes in the angle of attack. Other challenges, like the influence of the short-scale turbulence was considered by Peinke [53]. He showed that a small-scale atmospheric turbulence, which can results from wind gusts, can statistically arise the anomalous probabilities.

1.4. Objectives

The complex design of HAWT's that involves many interacting parameters is still revealing many challenges, in spite of many investigations. Design of wind turbine elements and trade of among them with considering of the site and the environmental influences, such as terrain, turbulence, wind shear and gusts are still a wide foggy area for researchers. In addition, the control strategies that are important to keep the wind turbine smoothly running in the designed rated performance and to protect it from damage under the influences of turbulence environment are crucial in designing HAWTs. Constraints arising from employed technology and the environmental conditions needs to be investigated too. The main questions associated with wind turbine operating under the turbulence are: What is the influence of the turbulence on the HAWT performance and what are the contributions of the turbulence characteristics.

The author claims that turbulence can impact the performance of the HAWT by different possible means. These are; interactions of turbulence different length and velocity scales with the blade surface boundary layer. This in turn, could effect the stall, the boundary layer thickness and separation. In addition, the higher turbulence levels might effect the tip vortex and hence the tip losses. Furthermore, high turbulence con-

1. Introduction and Outline

tent in the incoming wind can perform in the wake-surrounding exchange, and hence energy exchange. In the next section, the present adopted methodology to reach the main objective will be presented.

1.5. Methodology and Thesis Structure

In order to investigate the influence of the turbulence systematically, it is ought to design a laboratory-scale HAWT that efficiently perform in the LSTM wind tunnel. Its efficiency is in the range of the real-scale HAWTs. Thus, the obtained results are reliable and applicable for the different scales of HAWTs. For that reason, it is suggested firstly at developing a blade shape design optimization method constrained to the torque rotational speed characteristic of a generator, called Torque Matched Aerodynamic Shape Optimization (TMASO) method. It is based on modifying the turbine blade shape to extract the maximum possible power while it is coupled to any selected generator. In the design optimization process the Reynolds number is taken into account. Therefore, a low Reynolds-high lift airfoil is selected for the rotor blade section.

As a part of TMASO, a Torque-Matched Aerodynamic Performance Analysis Method (TMAPAM) which can predict the HAWT performance under various operation conditions has been developed. With this method the performance of any designed HAWT can be predicted with accounting of the generator's torque rotational speed characteristic.

In order to investigate different control strategies, the work involves an optimization method for pitch-control of the HAWT called Torque-Matched Pitch Control Optimization (TMPCO). It is constrained to the torque of the generator and aimed to keep the rated power constant by adjusting the blade pitch angle. Up to this point, a complete procedure for performance prediction, shape and pitch-control optimizations of new or existing wind turbines are developed in this work.

By using TMASO, a laboratory scale wind turbine was developed for wind tunnel tests. The tests are conducted in the closed loop wind tunnel of LSTM. The rotor blades are specially designed and optimized for this wind tunnel and the generator used. The performance of the produced HAWT model, which was analyzed with the developed method TMAPAM, is validated experimentally. This is done with a special designed setup that can measure the performance in the wind tunnel. whereas the torque needed for calculation is measured separately with a second experimental setup.

1. Introduction and Outline

In order to verify the applicability of the developed optimization methods for real scale HAWTs, they are applied to the National Renewable Energy Laboratory (NREL) different scales wind turbines of 10kW and 5MW. The NREL Unsteady Aerodynamics Experiment (UAE) phase-VI turbine of 10 kW is used here as a baseline because of its well-documented wind tunnel database that includes pressure distributions, separation boundary locations, drag data, and flow-visualization data [48]. The stall-regulated control strategy is considered in the analysis.

The prediction method of the stall-regulated turbine need modifications for the existing post-stall models, which is conducted successfully in the present work, and validated with available experimental data. In addition, the NREL 5MW wind turbine is used in the validation process as an example for a large scale pitch-controlled wind turbine to verify the accuracy of the method in predicting the operation performance for different scales and different control strategies HAWTs.

For further validation of the optimized turbine performance, a RANS numerical simulations with STAR-CCM+ were performed. The operating power coefficient at the design and off-design tip speed ratios is investigated and compared to experimental data. The flow over the wind turbine blade at different angles of attack is assessed in regards of separation and three dimensional flow on the blade. This provides a deep understanding of the flow behaviour over blade under different operating points.

Finally, to reach the main objective of the investigations, which is figuring out the performance of HAWT operating under the influences of turbulent conditions. This is done by exposing the produced efficient wind turbine model to turbulence with various energy content. It is done by using two static squared grids (Fine and Coarse). The developments of Taylor's micro scale (λ_g) and integral scale of the turbulence (L_g) in the flow direction at various Reynolds numbers based the mesh size of the grid (Re_M) are measured. Upstream and downstream turbulence intensities (TI) distributions are measured to give insight of the surrounding free-stream and turbine-wake interaction and how can different turbulence eddies scales influence the performance of the turbine. Performance measurements are conducted with and without turbulence grids and the results are compared.

Detailed measurements and analysis of the TI and the spectrum analysis are conducted. This provides insight of the different scales wake-surrounding interaction in both near and far regions. In addition, information of the tip vortex and penetrating of some scales into the wake region can be obtained.

1. Introduction and Outline

In chapter 2, basic theoretical background for the aerodynamics concepts, design theories, control strategies, optimization methods and the characterization of turbulence are presented. Chapter 3, introduces the numerical setup intended to be used in the numerical validation part.

The experimental Facilities such as, wind tunnel, grids and the instrumentations used in this work are presented in chapter 4.

Details of TMASO and TMPCO with their applications on the laboratory and real scales wind turbines are presented in chapter 5. Whereas, the performance analysis of TMAPAM is explained in chapter 6. In addition, performance analysis methods for different control strategies such as, the post-stall for the stall-regulated turbine and the pitch-control analysis are presented too.

In chapter 7, a numerical validation of the developed analysis methods is introduced with a special treatment for validation of the VS-VP turbine with the use of the overset mesh in STAR CCM+.

A detailed experimental investigations of the optimized wind turbine, which include measurements of the upwind and the downwind velocity distributions and turbulence generated by using different grids, are presented in chapter 8. Finally, conclusions and outlook are presented in chapter 9.

2
Theoretical Background

In this chapter, first the physical basics and rotor design fundamentals are presented; these are needed to to understand the design and analysis procedures and to interpret the validation results are presented. Second, the basic concepts of the control strategies governing the operation of the HAWT are introduced. Third, a brief background of optimization and comparison between the gradient-based, which is intended to be implemented in the present work, and the other optimization methods are explained. Finally, the turbulence characteristics and its basic concepts are presented.

2.1. Wind Power and Aerodynamics

2.1.1. Betz limit

A rotor of a wind turbine extracts the power from the wind P_w to produce mechanical energy by the rotor and then transform it into electric energy by a generator. P_w is composed of the mass flow rate, which is a function of air density ρ, velocity far in front of the rotor v_1 and rotor plane area A of the continuity equation of fluid mechanics [67].

Applying the continuity equation on a wind with the density ρ flowing with a velocity v_1 through a cross-section A, the mass flow rate $\dot{m}$ is

$$\dot{m} = \frac{dm}{dt} = \rho A v_1 \tag{2.1}$$

2. Theoretical Background

The power of the wind P_w is the derivative of its energy E in respect to the time t. With E being equal to $\frac{1}{2}mv_1^2$ the power of the wind is

$$P_w = \frac{dE}{dt} = \dot{E} = \frac{1}{2}\frac{dm}{dt}v_1^2 = \frac{1}{2}\dot{m}v_1^2 = \frac{1}{2}\rho A v_1^3 \qquad (2.2)$$

Betz's analysis assumes a control volume shown in Figure 2.1. In this figure, changes in wind velocity and pressure over the turbine plane are illustrated. For simplifying the analysis, it is assumed a control volume bounded by a side stream tube and two cross-sections, far upwind with wind velocity v_1 and far downwind with wind velocity v_3, whereas v is the velocity at the rotor plane. The rotor plane is represented by a uniform actuator disc.

The conversion from wind energy to mechanical energy is achieved by decelerating the air. The air cannot be slowed down completely, as a whole deceleration of the wind would choke the area for the replenishing air. Therefore, the equation of continuity would not be fulfilled. The other extreme, no change in velocity, gives no mechanical energy either. Thus, it is suggested that the highest power conversion lies between these boundaries, Betz [24]. To find out the optimum value of the axial velocity reduction, we will follow Betz' analysis procedures and assumption. The analysis is based on the Rankine-Froude momentum theory for an actuator disk theorem, [42] and the assumtions are as below,[67]:

- Homogeneous, incompressible, steady state fluid flow
- No frictional drag
- Infinite number of blades
- Non-rotating wake
- Static pressure far upstream and far downstream of rotor is equal to undistributed ambient static pressure

As shown in Figure 2.1, the wind turbine decelerates the wind velocity v_1, which leads to a pressure difference Δp that can be calculated by applying Bernolli's equation upwind and downwind the turbine, since there is no work done on either side of the turbine, and then subtracting, [47]:

$$\Delta p = \frac{1}{2}\rho(v_1^2 - v_3^2) \qquad (2.3)$$

In addition, another expression of the pressure difference across the turbine can be calculated from the change of axial momentum per one square meter of the rotor plane, [47]:

2. Theoretical Background

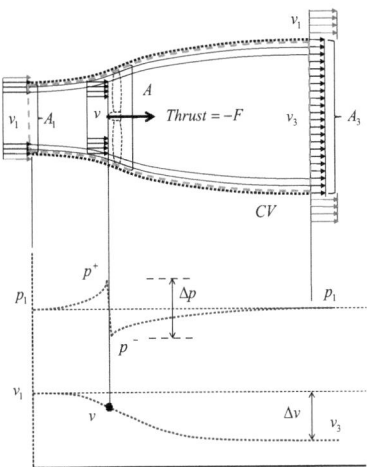

Figure 2.1.: Distribution of wind velocity and pressure over rotor plane.

$$\Delta p = \rho v (v_1 - v_3) \tag{2.4}$$

Equalizing 2.3 and 2.4 proves that the wind velocity at the rotor plane (or the velocity at the actuator disc) is the average of the upstream and downstream.

$$v = \frac{1}{2}(v_1 + v_3) \tag{2.5}$$

In the following the axial induction factor a is introduced, which defines the fractional decrease in wind velocity between the free stream and the rotor plane [67].

$$a = \frac{v_1 - v}{v_1} \tag{2.6}$$

From equations 2.6 and 2.5, v and v_3 can be described as below.

$$v = v_1(1 - a) \tag{2.7}$$

$$v_3 = v_1(1 - 2a) \tag{2.8}$$

The power of the turbine is composed of the volume flow rate and the pressure

2. Theoretical Background

difference (any expression for Δp can be applied; the one from Bernolli 2.3 or the axial momentum 2.4).

$$P = \dot{V}\Delta p = Av\Delta p = \frac{1}{2}\rho v(v_1^2 - v_3^2)A = \frac{1}{2}\rho v(v_1 + v_3)(v_1 - v_3)A \tag{2.9}$$

Substituting equations 2.7 and 2.8 gives:

$$P = \frac{1}{2}\rho A v_1^3 4a(1-a)^2 \tag{2.10}$$

Power Coefficient C_P of a wind turbine characterize its performance [67].

$$C_P = \frac{P}{P_w} = \frac{\frac{1}{2}\rho A v_1^3 4a(1-a)^2}{\frac{1}{2}\rho A v_1^3} = 4a(1-a)^2 \tag{2.11}$$

The maximum power coefficient ($C_{p,Betz} = \frac{16}{27} \approx 0.593$) is reached when a is equal to $\frac{1}{3}$, which is found by taking the derivative of C_p with respect to a and setting it equal to zero. According to equation 2.7, an ideal wind turbine reduces the flow at the rotor plane to $\frac{2}{3}$ of the wind velocity long in front of the rotor. Based on this condition, the maximum power of the wind can be extracted. The axial force (thrust) on the rotor can be calculated as:

$$F = A\Delta p = A\Delta p = \frac{1}{2}\rho(v_1^2 - v_3^2)A = \frac{1}{2}\rho(v_1 + v_3)(v_1 - v_3)A \tag{2.12}$$

following the same procedure to obtain:

$$F = \frac{1}{2}\rho A v_1^2 4a(1-a) \tag{2.13}$$

and the thrust coefficient C_F

$$C_F = \frac{F}{F_w} = \frac{\frac{1}{2}\rho A v_1^2 4a(1-a)}{\frac{1}{2}\rho A v_1^2} = 4a(1-a) \tag{2.14}$$

substitute the a=1/3 to obtain the $C_F, betz = 8/9$ ($C_{F,Betz} = \frac{8}{9} \approx 0.89$)

2.2. Airfoil Theory

2.2.1. Forces on airfoils

Wind turbine blade designs are based on airfoils. The cross section of a blade has the shape of an airfoil. The characteristics of an airfoil are mainly based on the leading edge radius, maximum thickness, thickness distribution and the mean camber line [67].

2. Theoretical Background

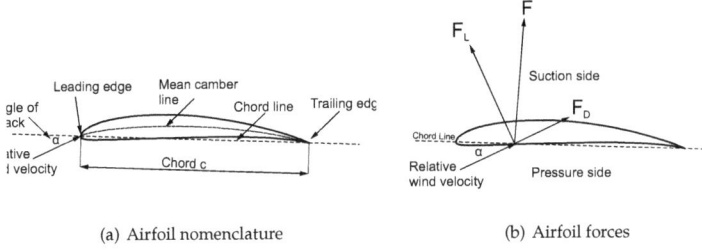

(a) Airfoil nomenclature

(b) Airfoil forces

Figure 2.2.: Airfoil aerodynamic characterizations

Nomenclature of an airfoil, which are illustrated in Figure 2.2.1(a) and its aerodynamic forces are depicted in Figure 2.2.1(b). These forces, which are shown in figure, have an influence on the power extraction of a wind turbine. The mean camber line has its location halfway between the upper and lower surfaces of the airfoil. The leading edge is the most forward and the trailing edge is in the most rearward position on the mean camber. The chord line connects both edges and the distance between them is named as chord length c. The camber describes the distance between the chord line and mean camber line along the chord. Thickness of an airfoil is defined as the distance between the upper and lower surfaces [67]. In Figure 2.2.1(b), the lift force F_L and drag force F_D of an airfoil are illustrated. They are induced by the air flow, which hits the blade at an angle of attack named α. It defines the angle between the chord line and relative wind velocity w. The relative wind velocity is explained in section Design theory of Betz.

The lift force is perpendicular to the relative wind velocity and it is a result of pressure differences between the lower and upper airfoil surfaces. On the contrary, the drag force is parallel to the relative wind velocity. It develops because of viscous friction forces at the surface of the airfoil and pressure differences between the lower and upper airfoil surfaces, which face toward and away from the oncoming flow [67]. In airfoils analysis, there are many non-dimensional numbers used for defining the flow characteristics. The most important one, which considers fluid flow conditions and viscosity, is the Reynolds number, Re.

$$Re = \frac{\rho v_1 L}{\mu} = \frac{Inertial\ force}{Viscous\ force} \tag{2.15}$$

where, ρ is the air density, μ is the air dynamic viscosity, L is the length and v_1 is the air velocity. For wind turbines applications to have more useful Re number, the last two parameters in the numerator could be replaced by the relative velocity w and the

2. Theoretical Background

chord length c, respectively. The result is the so-called chord Reynolds number $Re_{,c}$. Additional important non-dimensional numbers are forces coefficients (lift and drag). The lift and drag forces are calculated as:

$$F_L = C_L \frac{1}{2} \rho w^2 (bc) \tag{2.16}$$

$$F_D = C_D \frac{1}{2} \rho w^2 (bc) \tag{2.17}$$

where b is the width of the blade, C_L is the coefficient of lift and C_D is the coefficient of drag. Both coefficients are dependent on both α and Re. The stall angle is typically in the range of 15° to 20°, where the air is no longer attached to the blade [47]. This results in a decrease of C_L. The Glide ratio, which is defined as $GR = C_L/C_D$, is an important parameter that is decisive for the performance of wind turbines. It is the aim to achieve a maximum GR which is mainly associated with the design angle of attack (α_d) in the range of 5° to 10° [47].

2.2.2. Drag coefficient components

The drag coefficient is composed of two components: induced drag coefficient C_{Di} and zero-drag coefficient C_{D0}. C_{Di} is associated with the dependency on the lift, hence the angle of attack and it is composed of vortex drag and wave drag. Vortex drag relates to vortices around the tip of the blade. Wave drag considers air compressibility effects, which can be neglected in subsonic flows. C_{D0} includes all types of drag, which are not dependent on the production of lift as profile drag. Profile drag consists of skin friction drag, which results from viscous shearing stresses, and form drag that results from pressure differences. The induced drag coefficient and zero-drag coefficient are both dependent on wind velocity, air density, reference area. Moreover, the induced drag coefficient is dependent on the lift coefficient and the zero-drag coefficient is a function of the external shape of the components. The drag coefficient is calculated as, [78]:

$$C_D = C_{D0} + C_{Di} = C_{D0} + \frac{C_L^2}{\pi e\, AR} \tag{2.18}$$

where e is the blade span efficiency. Its value is between 0 and 1. An ellipse blade shape relates to a blade span efficiency of $e = 1$. AR is the aspect ratio, it is defined as blade length divided by chord length [78].

2.2.3. Separation and Stall

The boundary layer that develops over the pressure and suction surfaces due to friction, has high influence on the flow over the airfoil [81]. Especially in the suction side, the boundary layer is subject to separated flows. The flow on the convex suction surface is at first accelerated and the pressure thus decreases. At some point along the blade, the flow is decelerated due to the airfoil geometry, whereby the pressure increases again. If the kinetic energy of a fluid particle in the boundary layer is too low to overcome this pressure increase a reversed flow in the boundary layer occurs. This reversed flow causes flow separation. This flow separation influences the drag force on the airfoil. It is increased due to the added form drag. This phenomenon mainly is influenced by the airfoil shape, Reynolds number and angle of attack, whereby low Reynolds numbers and high angles of attack induce separation [69]. When angle of attack increases, there is, at some point, a massive flow separation leading to a large increase in drag.

2.3. Design Theories

In this section, design theories of airfoils by Betz, Schmitz and Blade element momentum (BEM) are explained. The objective of all theories is to produce a turbine blade by calculating the chord length $c(r)$ and the twist angle $\beta(r)$ distributions. In contrast to the design theory of Betz, Schmitz and BEM consider the rotation of the wake. Both theories neglect the drag coefficient in their calculations [67].

2.3.1. Design theory of Betz

Theory of Betz is mainly based on combining the momentum with the blade tangential component force with the assumptions of optimum induction factor of $a = 1/3$ and no wake rotation $a' = 0$. Thus, it determines an ideal blade shape or 'Betz optimum rotor'. In Figure 2.3, velocities and angles, which are used for the theory of Betz, are illustrated.

All angles are function of radius, and are listed as:

- $\alpha_{(r)}$ = angle of attack
- $\beta_{(r)}$ = summation of tip pitch angle β_o and twist angle $\beta_{t(r)}$.
- $\varphi_{(r)}$ = angle of relative wind to rotor plane

The relative wind velocity w is composed of tip speed u and wind velocity v as below:

$$w^2 = v^2 + u^2 \tag{2.19}$$

The tip speed u is calculated as:

$$u = wr \tag{2.20}$$

2. Theoretical Background

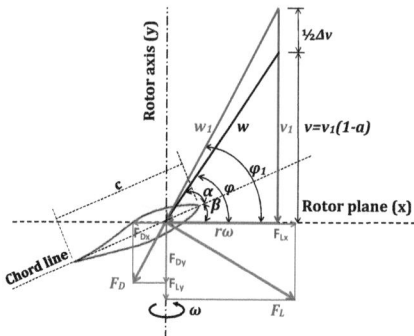

Figure 2.3.: Angles and velocities on the airfoil before and at the rotor plane (Betz triangles).

where ω is the angular speed ($\omega = 2\pi n/60 \ [rad/s]$) and n is the rotational speed of the rotor $[RPM]$.

An important non-dimensional number for further calculations in the design theories is the tip speed ratio λ [47]:

$$\lambda = \frac{\omega R}{v_1} \quad or, \ the \ local \ speed \ ratio \quad \lambda_{(r)} = \frac{\omega r}{v_1} \tag{2.21}$$

It is composed of the tip speed divided by the wind velocity far in front of the rotor.

From the velocity triangle of Figure 2.3 and using of equation 2.7 with Betz's optimum condition of $a = \frac{1}{3}$ and equation 2.21 leads to:

$$\varphi(r) = tan^{-1}\frac{v}{u} = tan^{-1}\frac{2R}{3\lambda r} \tag{2.22}$$

and

$$\beta(r)_{Betz} = tan^{-1}\frac{2R}{3\lambda r} - \alpha_d \tag{2.23}$$

In order to calculate the chord length $c(r)_{Betz}$, the width b of the blade in equations 2.16 and 2.17 is replaced by the thickness dr for one blade element in the distance r from the rotor axis, Figure 2.4.

The lift and drag forces of the element of area $A_b = cdr$ are :

$$dF_L = C_L \frac{1}{2}\rho w^2 cdr \tag{2.24}$$

2. Theoretical Background

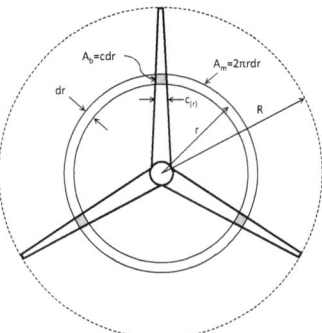

Figure 2.4.: Rotor sections.

$$dF_D = C_D \frac{1}{2}\rho w^2 c dr \tag{2.25}$$

Considering one blade element, the tangential component of the forces is dF_x, and the normal component of the forces (thrust) is dF_y of the rotor, which are demonstratively shown in Figure 2.3, are calculated as:

$$dF_x = \frac{1}{2}\rho w^2 c dr (C_L \sin(\varphi) - C_D \cos(\varphi)) \tag{2.26}$$

$$dF_y = \frac{1}{2}\rho w^2 c dr (C_L \cos(\varphi) + C_D \sin(\varphi)) \tag{2.27}$$

Assuming design conditions with maximum GR, C_L is much greater than C_D. Hence, the resultant power of the blade element, with number of blades B, is shortened to:

$$dP = BdF_x r\omega = B\frac{1}{2}\rho w^2 c dr C_L \sin(\varphi) r\omega \tag{2.28}$$

which represents the power extracted by the element of the rotor blades.

After Betz's theory, the loss of the wind power passing through the rotor was calculated from the energy equation or momentum as in equation 2.10 for an area of a ring element A_m, Figure 2.4 with the optimum condition of $a = \frac{1}{3}$:

$$dP = \frac{16}{27}\frac{1}{2}\rho v_1^3 (2\pi r dr) \tag{2.29}$$

The chord length is found by equating equations 2.28 and 2.29, with using of relations from the Betz velocities triangle of Figure 2.3 $v_1 = \frac{3}{2}w\sin(\varphi)$ and $u = w\cos(\varphi)$ as:

18

2. Theoretical Background

$$c(r) = \frac{16\pi R}{9BC_{L,d}} \frac{1}{\lambda \sqrt{\lambda^2 (\frac{r}{R})^2 + \frac{4}{9}}} \tag{2.30}$$

where $C_{L,d}$ is the lift coefficient at the design angle of attack.

The same expression for the chord can be found by equating the blade normal force equation 2.27 with the momentum axial force equationer 2.13.

2.3.2. Design theory of Schmitz

The power is calculated as:

$$P = T\omega \tag{2.31}$$

As a result of the consideration of the wake, it can be theoretically shown that the tip speed u in the rotor plane is half of the total change. This defines the tangential induction factor a':

$$u = r\omega + \frac{1}{2}\Delta u$$
$$= r\omega(1 + a') \tag{2.32}$$

where, $a' = \omega'/2\omega$.

The following equations result from Figure 2.5, where subscript number 1 is far in front of the rotor plane, without subscript is in the rotor plane [47].

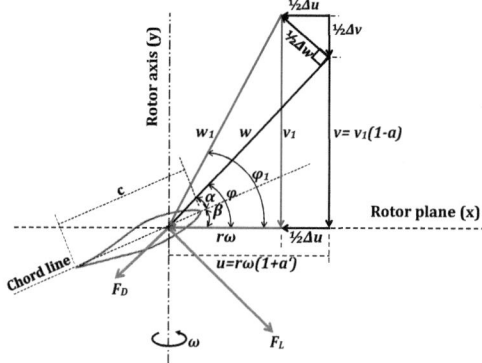

Figure 2.5.: Velocities triangles before and at the rotor plane.

$$w = w_1 \cos(\varphi_1 - \varphi) \tag{2.33}$$

2. Theoretical Background

$$v = w_1 \cos(\varphi_1 - \varphi) \sin \varphi \tag{2.34}$$

$$\Delta w = 2w_1 \sin(\varphi_1 - \varphi) \tag{2.35}$$

In order to calculate the power, the mass flow ($\dot{m} = 2\rho\pi r \, dr \, v$) is needed as well. The linear momentum transport along the direction of the relative velocity change Δw results in the following lift force on the blade element:

$$dF_L = \Delta w \, d\dot{m} \tag{2.36}$$

With neglecting the drag force, Δw vector will be in the direction of the dF_L vector. Hence, the power extracted from changing of relative velocity is,

$$\begin{aligned} dP &= dT\omega \\ &= dF_x r\omega \\ &= dF_L \sin(\varphi) r\omega \\ &= \Delta w \, d\dot{m} \, sin(\varphi) \, r \, \omega \\ &= 2w_1 \sin(\varphi_1 - \varphi) \left[(2\rho\pi r dr) \, w_1 \cos(\varphi_1 - \varphi) \sin(\varphi)\right] \sin(\varphi) \, r\omega \end{aligned} \tag{2.37}$$

Equation 2.37 is differentiated with respect to φ and solved by equalizing the equation with zero to reach the angle of $\varphi = \varphi_x$, which predicts maximum power [42]:

$$\varphi_x = \frac{2}{3}\varphi_1 \tag{2.38}$$

By calculating φ_1 from Figure 2.5 and using equation 2.38, φ_x can be written as:

$$\varphi_x = \frac{2}{3} \arctan \frac{R}{\lambda \, r} = \frac{2}{3} \arctan \frac{1}{\lambda_r} \tag{2.39}$$

The relationship $\alpha + \beta = \varphi$ is valid for Schmitz as it is in Betz. As a result of the summation of the tip pitch angle β_o and twist angle distribution β_t of Schmitz, the pitch angle distribution is calculated as [42]:

$$\begin{aligned} \beta(r)_{Schmitz} &= \varphi_x - \alpha_d \\ &= \frac{2}{3} \arctan \frac{R}{\lambda \, r} - \alpha_d \end{aligned} \tag{2.40}$$

In order to calculate the chord length distribution of the Schmitz equation 2.34, 2.35 and 2.38 are used for calculating dF_L:

2. Theoretical Background

$$dF_L = 2w_1^2 2\rho\pi r dr 2\sin^2\left(\frac{\varphi_1}{3}\right)\cos^2\left(\frac{\varphi_1}{3}\right) \tag{2.41}$$

noting that $sin(2x) = sin(x)cos(x)$. By applying the airfoil theory and using of equation 2.33, dF_L is calculated as:

$$dF_L = \frac{1}{2}\rho w_1^2 B c dr C_L \cos^2\left(\frac{\varphi_1}{3}\right) \tag{2.42}$$

The chord length distribution of Schmitz results from equating equations 2.41 and 2.42:

$$c(r)_{Schmitz} = \frac{1}{B}\frac{16\pi r}{C_L}\sin^2\left(\frac{1}{3}\arctan\left(\frac{R}{\lambda r}\right)\right) \tag{2.43}$$

2.3.3. Blade Element Momentum (BEM) theory

The BEM method is based on dividing the rotor into a number of radius direction rings. Then it is possible to calculate axial force F_x and power P for each ring element of the rotor for different wind velocities by iteration of the axial and tangential induction factors [67]. From Figure 2.5, the following equations can be obtained:

$$\alpha = \varphi - \beta \tag{2.44}$$

$$\tan(\varphi) = \frac{1-a}{1+a'}\frac{1}{\lambda_r}$$
$$= \frac{1-a}{1+a'}\frac{v_1}{r\omega} \tag{2.45}$$

According to equation 2.44 and 2.45, the angle of attack increases from the tip to the hub and with increasing wind velocity v_1. By using laws of axial and angular momentum, thrust dF and torque dT of the rotor of one ring element of area A_m (see Figure 2.4) are calculated as:

$$dF = 2\pi r \rho v (v_1 - v_3) dr \tag{2.46}$$

$$dT = d\dot{m}(u_3' - u_1')r = 2\pi r^2 \rho v u_3' dr \tag{2.47}$$

where u_3' is the wake speed far behind the rotor plane and since these is no upwind wake rotation $u_1' = 0$, so it is exactly Δu in equation 2.32. $u_3' = \Delta u = 2a'\omega r$. The far downstream axial velocity v_3 can be written in terms of a and v_1 by using of equation 2.8.

From the blade element theory, it is possible to calculate the same quantities by proceeding with equations 2.26 and 2.27, which are already calculated for a ring blade

2. Theoretical Background

element of area A_b shown in Figure 2.4, and multiply them with number of blades B,

$$dF = dF_y B = \frac{1}{2}\rho w^2 c\, dr\, C_y B \tag{2.48}$$

$$dT = dF_x\, rB = \frac{1}{2}\rho w^2 c\, dr\, C_x\, rB \tag{2.49}$$

where, $C_y = C_L \cos(\varphi) + C_D \sin(\varphi)$ and $C_x = C_L \sin(\varphi) - C_D \cos(\varphi)$. For simplicity, it is possible to neglect the drag thus dropping the second term from both equations.

Combination of thrust equations (2.48 and 2.46) and torque equations (2.49 and 2.47) results in [67, 79] using of the relation $w = v_1(1-a)/sin(\varphi) = \omega r(1+a')/cos(\varphi)$ from Figure 2.5,

$$\frac{a}{a-1} = \frac{cBC_y}{8\pi r \sin^2(\varphi)} \tag{2.50}$$

$$\frac{a'}{a'+1} = \frac{cBC_x}{8\pi r \sin(\varphi)\cos(\varphi)} \tag{2.51}$$

Defining the solidity ratio as $\sigma = \frac{cB}{2\pi r}$ and consider Prandtl's tip loss correction factor F, which is calculated as [47]:

$$F = \left(\frac{2}{\pi}\right) \cos^{-1}\left[\exp\left(-\frac{(B/2)(1-(r/R))}{(r/R)sin\varphi}\right)\right] \tag{2.52}$$

The axial and tangential induction factors can be calculated as:

$$a = \frac{1}{\frac{4F sin^2\varphi}{\sigma C_y} + 1} \tag{2.53}$$

$$a' = \frac{1}{\frac{4F sin\varphi cos\varphi}{\sigma C_x} - 1} \tag{2.54}$$

Prandtl's tip loss correction factor is needed due to the pressure difference between the suction and the pressure sides of the blade that makes the air tend to flow around the tip, reducing lift force and hence power production near the tip [67]. If the value of a becomes greater than $a_c = 0.2$, simple BEM theory breaks down. Therefore, the following equation 2.55 has to replace equation 2.53 [47].

$$a = \frac{1}{2}(2 + K(1-2a_c) - \sqrt{(K(1-2a_c)+2)^2 + 4(Ka_c^2 - 1)} \tag{2.55}$$

where,

$$K = \frac{4F sin^2\varphi}{\sigma C_y} \tag{2.56}$$

2. Theoretical Background

After defining the equations to calculate a and a', the iteration can begin to calculate F, T and P for a ring element as [79]:

1. Initial values of a and a' (normally between 0 and 0.5) are assumed.

2. For a given design λ, φ is calculated by equation 2.45 and accordingly α by equation 2.44 with a given setting of blade pitch distribution $\beta(r)$ (note that $\beta(r)$ can be found either with Betz or Schmitz theories).

3. Calculation of C_L and C_D by blade profile data or by sufficient prediction models.

4. Calculation of a and a' by equation 2.53 and 2.54. Or if a is greater than 0.2 then it is calculated by equation 2.55.

5. Iteration continues until the values of a and a' converge (the error less than 1%).

After evaluating a and a', axial force on the turbine can be calculated as [67, 79],

$$F = \int_0^R dF = \int_0^R 4a(1-a)\rho v_1^2 \pi r dr \qquad (2.57)$$

and the Torque,

$$T = \int_0^R dT = \int_0^R 4a'(1-a)\rho v_1 \pi r^3 \omega dr \qquad (2.58)$$

Hence, the Power P is calculated by multiplying the Torque with the angular speed equation 2.31. The power coefficient C_p results from dividing power by wind power according to equation 2.11.

2.3.4. Lifting-Surface Prescribed-Wake Performance Prediction Method

Beside the aforementioned theories, there is another one, which is only performance-prediction method named Lifting-Surface Prescribed-Wake (LSWT). While BEM theory is based on an iteration of a and a to predict the power, the LSWT method determines the blades angle of attack distribution with an iterative process. In contrast to BEM theory, the LSWT method "accounts for the induced effects of the blade configuration and those from the span-wise distribution of trailing vorticity in calculating the angle of attack" [71].

The method of LSWT predicts α by using blade pressure measurement in conjunction with a free wake vortex [80]. At first an initial span-wise angle of attack distribution is assumed. Meanwhile, normal force coefficient (C_n) and tangential force coefficient (C_t) are calculated by measured pressure of pressure tap rows along the blade

2. Theoretical Background

($\Delta p = p_o p_u$). Since, it is possible to have the below relations,

$$C_n = C_L \cos\alpha + C_D \sin\alpha \qquad (2.59)$$

$$C_t = C_L \sin\alpha - C_D \cos\alpha \qquad (2.60)$$

hence, it is possible to find out C_L and C_D in term of C_n, C_t and α.

By means of the Kutta-Joukowski theorem, the bound circulation distribution at the blades is predicted and prescribed to the free wake vortex model to generate the free vortical wake [80]. Knowing the bound circulation, local wind velocity (v_l) can be calculated and used for the prediction of a new angle of attack distribution. This iteration is continued until convergence between the assumed and the predicted α is achieved [80].

2.3.5. Efficiency of the wind turbine

Efficiency of a wind turbine is defined as the ratio of rotor power by maximum power. P_{max} has a power coefficient of $16/27$ according to Betz in equation 2.29.

$$\eta_{rotor} = \frac{P_{rotor}}{P_{max}} = \frac{P_{rotor}}{\frac{16}{27}\frac{1}{2}\rho v_1^3 A} \qquad (2.61)$$

The rotor efficiency is divided into wake losses, tip losses and profile losses, [47]:

$$\eta_{rotor} = \eta_{wake}\,\eta_{tip}\,\eta_{profile} \qquad (2.62)$$

The wake losses can be calculated as the difference between the power coefficient of Schmitz and Betz due to the swirl losses and rotation of the wake, which are taken into account by Schmitz:

$$\eta_{wake} = \frac{C_{p,Schmitz}}{C_{p,Betz}} \qquad (2.63)$$

The tip losses result from a pressure difference around the blade tip. There is a high negative pressure above the blade and a little positive pressure under the blade. This pressure difference leads to a by-pass flow near the tip from the high pressure side to the low pressure side, which reduces the pressure difference at the tip and hence power. Beside Prandtl's tip loss correction factor, the tip losses can be considered by the following equation:

$$\eta_{tip} = \left(1 - \frac{0.92}{B\sqrt{\lambda^2 + 4/9}}\right)^2 \qquad (2.64)$$

The profile losses arise from the fact that the drag coefficient is significant by consideration of real wind turbines. A high drag coefficient leads to a decrease in power prediction:

$$\eta_{profile}(r) = 1 - \frac{3r\lambda}{2R\ GR} \tag{2.65}$$

2.4. Wind Turbines Control Strategies

Objectives to control wind turbines are mostly to maximize energy capture by taking into account the operation conditions such as cut-in, rated and cut-out wind velocity and to prevent the wind turbine from excessive dynamic mechanical loads to guarantee low maintenance costs, safe operation and long structural life [3, 6, 2].

To maximize energy capture, operating at maximum power coefficient is compulsive until rated power is reached. The power coefficient of a wind turbine dependent on the tip speed ratio and the torque coefficient:

$$C_P = C_T \lambda \tag{2.66}$$

The wind turbine is normally designed for one optimum tip speed ratio λ_0, which relates to the maximum power coefficient $C_{P,max}$. According to equations 2.11, 2.31 and 2.66 the torque of the rotor is calculated as:

$$T = \frac{P_w C_T \lambda}{\omega} \tag{2.67}$$

Equation 2.67 can be used to generate Figure 2.6. The torque for $C_{P,max}$ is calculated by [27]:

$$T = K\omega^2 \tag{2.68}$$

where K is a constant, which is calculated as:

$$K = \frac{1}{2\lambda_o^3} \rho \pi R^5 C_{p,max} \tag{2.69}$$

The maximum torque is found by inserting the power coefficient of Betz in equation 2.69. For NREL phase-VI the rotor rated power is $10kW$, by reaching $\omega = 7.5 rad/s$ rated torque is obtained from $v_1 = v_N = 9.4 m/s$. The important region for control strategies is up until the angular speed reaches $7.5 rad/s$ and hence rated power. This region is marked in figure 2.6.

In the following four different control strategies are explained:

- fixed-speed, fixed-pitch (FS-FP)
- variable-speed, fixed-pitch (VS-FP)

2. Theoretical Background

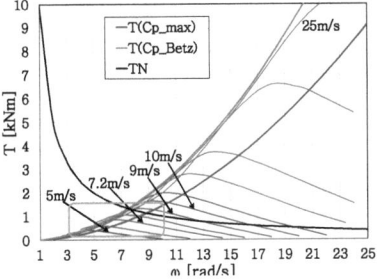

Figure 2.6.: Locus of torques of $C_{P,max}$, $C_{P,Betz}$ and of rated power over angular speed.

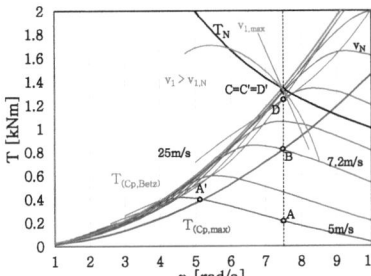

Figure 2.7.: Different control strategies

- fixed-speed, variable-pitch (FS-VP)
- variable-speed, variable-pitch (VS-VP)

For NREL phase-VI the cut-in wind velocity is $v_1 = 5m/s$, optimum wind velocity is $v_{1,o} = 7.2m/s$, rated torque T_N is reached at v_N and cut-out wind velocity is $v_1 = 25m/s$. The FS-FP is a passive control strategy. The angular speed remains fixed at ω_o regardless of which incoming wind velocity to couple an asynchronous generator directly to the power grid. The wind turbine is regulated by considering the appearance of stall. According to section Blade Element Momentum (BEM) theory and Forces on airfoils, the angle of attack increases with an increasing wind velocity. After reaching the stall angle, the air is no longer attached to the blade, which reduces the aerodynamic forces and, consequently, power production. In Figures 2.7 and 2.8(a), the control strategy of FS-FP is illustrated. Because of a constant angular speed, only at $v_{1,o} = 7.2m/s$ (B) the torque of the wind turbine reaches $C_{p,max}$. Hence, the power coefficient is less than its optimum value between (A) and (B) and thus power prediction. After reaching the rated torque at $v_1 = 15m/s$ (C), the torque is reduced by the appearance of stall, accordingly power production and power coefficient. Summing up, FS-FP control strategy has a poor power regulation, which is not able to keep the power production constant because of constrained operation conditions [3, 6, 2].

The control strategy of VS-FP is shown in same figures 2.7 and 2.8(a), adjusts the angular speed relative to the wind velocity by power electronics controlling the synchronous speed of the generator between the cut-in wind velocity (A') to $v_{1,o} = 7.2m/s$ (B). This modification results in an optimum tip speed ratio λ_o between (A') and (B), hence maximum power coefficient and maximum power production. In this control strategy, the generator is not connected to the power grid until $v_{1,o}$ (B) is reached, thus

2. Theoretical Background

the generator's drive-train is free to rotate independently of grid frequency. As a result the control strategy of VS-FP has in contrast to FS-FP between (A') and (B) a maximized power coefficient and agrees with the typical power curve until $v_{1,o}$ is reached.

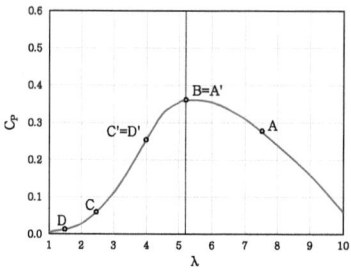

(a) Power coefficient versus tip speed ratio from cut-in to cut-out wind velocity

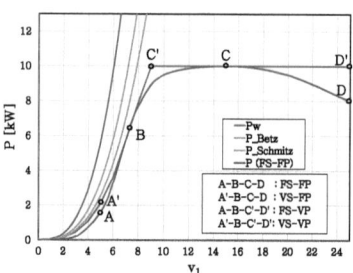

(b) Power extraction versus wind speed

Figure 2.8.: Different control strategies

In opposition to FS-FP and VS-FP the control strategies of FS-VP and VS-VP adjust the pitch angle of the blade after rated wind velocity is reached. It is done by turning the blade around its own axis to reduce the angle of attack while keeping the tip speed and accordingly angular speed constant. Reduction in α leads to a decrease in the aerodynamic forces, which facilitates constant power production. In contrast to stall-regulation, a pitch-regulated turbine has greater capital costs and more maintenance due to more moving parts.

Control strategy of FS-VP is illustrated in Figures 2.7 and 2.8(a). Angular speed remains constant at ω_o between (A) and (B) similar to the FS-FP control strategy. Due to the variable pitch control angle the torque is modified above the rated wind velocity $v_{1,max}$ to match the rated torque (C'=D'). Constant rated torque results in constant rated power and constant power coefficient between (C') and (D'). Furthermore, the power coefficient at (C'=D') of this control strategy is higher than the power coefficient at (C) of the control strategy of FS-FP because rated power is reached at a lower wind velocity $v_N \approx 9.4[m/s]$. Summing up, FS-VP control strategy has a poor power regulation until $v_{1,o}$ (B) is reached, though it reaches a maximized power between (C') and (D'). By reaching the cut-out wind velocity the pitch adjustment cannot limit the rotational speed any longer.

Control strategy of VS-VP integrates both VS-FP and FS-VP, as shown in Figures 2.7 and 2.8 (a), where angular speed is adjusted relative to the wind velocity between (A') and (B) to maximize energy capture and power coefficient. Above $v_{1,o}$, the angular speed remains constant at ω_o. By reaching the rated wind velocity $v_N \approx 9.4 m/s$, the pitch angle of the blade is adjusted to provide a constant power extraction (C'=D'). Theoretically, this control strategy reaches the typical power curve and operates most efficiently. Figure 2.8(b) shows all control strategies power-velocity curve, whereas VS-VP has the maximum power extracted over the range of the oncoming wind speed.

2.5. Optimization

Optimization methods are generally divided into many methods depend on the search problem, which could be either discrete or continuous. For the discrete (or combinatorial) problems, the meta-heuristic methods are commonly used. The problem is combinatorial if it consists of finding the optimal from a number of candidate options. The meta-heuristic methods are iterative methods that use and drive heuristics by combining different strategies for exploring and exploiting, experience accumulated, the search space to find out the best from many solutions, which might end up with near-optimal solution. They include different levels of learning process search. These methods are good solutions for a large, complex and non-linear search space because of their exploration capabilities [15]. Due to their search strategies they can easily skip out from the local minima. The search strategy of meta-heuristics methods are different. The best known are the evolutionary algorithms, which depend on stochastic searches such as swarm intelligence (SI) and genetic algorithm (GA) methods. In contrast, for the continuing problems, which can be solved mathematically, the deterministic optimizations are preferred, since they can perform more robustly and provide an exact solution.

2.5.1. Evolutionary Methods

The swarm intelligences (SI) methods are optimization techniques that are based on the movement and intelligence of swarm, examples of SI are particle swarm optimization (PSO) and ant colony optimization (ACO). These techniques use swarm behaviour to solve the problem, i.e. they use the concept of group intelligence to enhance the individual intelligence [76]. It uses a number of agents (particles) that form a swarm moving over the search space and looking for the best solution. In its procedure, each particle is searching for the optimum where it is moving with its own velocity. Each particle remembers its path, velocity and position, in which it had the best result so far, which is called the personal best p_{best}. This is not always the best global solution, thus, the particle needs informations from the neighbourhood particles to figure out where to search. Hence, the particles in the swarm exchanging the information about what they have discovered in their own visited paths. These information help each particle to adjust its velocity and move gradually towards the global best g_{best}. Accordingly, each particle tries to modify its search position s^k to better position s^{k+1} using current positions s^k, current velocities v^k, the distance between s^k and $s_{p_{best}}$ and the distance between s^k and $s_{g_{best}}$. Then, each particle uses these information to accelerate toward its p_{best} and the g_{best} locations, with a random weighted acceleration at each iteration.

The genetic algorithms (GA) is another important example of a heuristic, randomized, population-based search method that derived from natural selection and evolution. The general form is,

$$Minimize\ f(x),\ Subjected\ to\ x \qquad (2.70)$$

The decision variables x are encoded with an equal length of strings and/or numbers, which represent a population P in the search set (ω). Each individual within P should has an encoded x which is called "chromosomes", whereas the encoding process is called the "representation scheme". There are many ways to encode elements of solutions including binary, value, and tree encodings. For each chromosome x there is a corresponding function $f(x)$. f represents both the original objective function and the fitness measure of x. An appropriate representation for initializing the first population $P(0)$ is done by random selection of a set of chromosomes. The selection Imitates nature, where it selects the elitisms chromosomes into a mating pool $M(k)$ with the same number of elements as $P(k)$ by means of probabilities proportional to their fitness. The next step is the finding and evaluation of the next generations of $P(k+1)$ from $M(k)$. Usually, this is done by applying operations of crossover and mutation to the previous population. During each iteration, for each x_k the fitness function $f(x_k)$ is evaluated. Crossover and mutation, which are operators based on reproduction, are used to create the next generation of the population. Crossover combines some elements of solutions in the current generation to create a member of the next generation. Mutation systematically changes elements of a solution from the current generation in order to create a member of the next generation. Crossover and mutation are the evolution step that accomplish exploration of the search space by creating diversity in the members of the next generation.

2.5.2. Gradient-Based Optimization

The deterministic gradient-based optimization is preferred here over all other methods including the stochastic evolutionary, since the objective function is differentiable. Hence, the deterministic gradient-based method can perform more robustly and nor-

2. Theoretical Background

mally provides an exact solution. The governing equations explained in section Design Theories show that the objective function (the power coefficient) is non-linear, hence, the minimizing takes the general form,

$$min \ f(x) \tag{2.71}$$

when subjected to a linear constraint,

$$min \ f(x), \ such \ that \ Ax = b \tag{2.72}$$

and box-constraints (bounds)

$$min \ f(x) \ such \ that \ lb \leq x \leq ub \tag{2.73}$$

where ub and lb are vectors of upper and lower bounds, respectively. The box constraints problems ensure a sequence of strictly feasible points. x is the dependent variable vector, its transpose is: $x^T = [x_1, x_2, \cdots, x_n]$. Hence, the gradient vector of the function $f(x)$ is the partial derivative with respect to each of its independent variables is denoted as $\nabla f(x) \equiv g(x)$. For any differentiable multi-variable function, the second order partial derivative can be represented by a square symmetric matrix called Hessian matrix $\nabla^2 f(x) \equiv H(x)$. The sign of H decides the minima of the point x^*. This condition is called the second order sufficient condition, which implies that the Hessian matrix H has positive eigenvalues λ_i.

In non-linear constrained optimization, which takes the form:

$$\begin{aligned} & minimize \ f(x) \\ & subjected \ to \ h_i(x) = 0, \quad i = 1...m \\ & g_j(x) \leq 0, \quad j = 1...p \end{aligned} \tag{2.74}$$

The general aim is to transform the problem into an easier sub-problem that can then be solved and used as the basis of an iterative process. These methods are based mainly on the solution of the Karush-Kuhn-Tucker (KKT) equations. KKT are first order necessary conditions for solving non-linear programming (NLP) subjected to inequality constraints. The inequality constraint is active if $g_j(x) = 0$. The index for defining the activation within the feasible region in called $I(x)$. The point x^* will be a regular and

2. Theoretical Background

local optimum of the NLP if there is a vector μ_j such that:

$$\begin{aligned}
\nabla f(x^*) + \mu_j^T \nabla g_j(x^*) &= 0^T \\
g_j(x^*) &\leq 0 \\
\mu_j &\geq 0 \\
\mu_j^T g_j(x^*) &= 0
\end{aligned} \tag{2.75}$$

This implies that the gradient equals to the negative linear combination of the gradients of the binding active constraints, since the only active case is when $\mu_j \geq 0$, hence $g_j(x^*) = 0$. This option is called the active-set, which implies that the cancelling of the gradients between f and the active constraints $g_j(x)$ at the optimum point x^* is the first of equation of the KKT equations set 2.75. In order to cancel the gradients, only the equality multipliers are necessary. Hence, only equality and active inequality constraints are needed in this cancelling operation, so, the inactive constraints must excluded from the solution, which implies setting their associated multipliers to zero. Its advantage is only to determines constraints that influence the final result of optimization. The Sequential quadratic programming (SQP) iterative method of optimization are also employed in the current optimization. SQP methods are used on non-linear problems where the function and the constraints are double continuously differentiable. At each iterated point x_k, SQP defines an appropriate search direction d_k as a solution to the QP sub-problem. The solution is used for calculation of the new iteration as:

$$x_{k+1} = x_k + \alpha_k d_k \tag{2.76}$$

where the step length α_k can be calculated by means of line search methods to achieve a sufficient decrease in the merit function.

$$\alpha_k = arg\ min_{|\alpha \geq 0} f(x_k - \alpha g_k) \tag{2.77}$$

Non-linear constrained algorithm has a special function in MATLAB toolbox called $fmincon$, which is a gradient-based framework. It finds the minimum of a constrained

2. Theoretical Background

nonlinear multivariable function and takes the general form [13]:

$$\begin{aligned}
&\min f(x), \; subjected \; to \\
&c(x) \leq 0, \quad (nonlinear \; inequalities) \\
&ceq(x) = 0, \quad (nonlinear \; equalities) \\
&Ax \leq b, \quad (linear \; inequalities) \\
&Aeq \, x = beq, \quad (linear \; equalities) \\
&lb \leq x \leq ub, \quad (lower \; and \; upper \; bounds)
\end{aligned}$$

(2.78)

where x, b, beq, lb, and ub are vectors, A and Aeq are matrices, c(x) and ceq(x) are functions that return vectors, and f(x) is a function that returns a scalar. f(x), c(x), and ceq(x) can be nonlinear functions.

2.5.3. Evolutionary versus Gradient

Traditional uses of stochastic methods leverage the fact that they explore multiple areas of the search space to find a global minimum. For instance, through the use of the crossover operator, GA are particularly strong at combining the best features from different solutions to find one global solution. In addition, stochastic methods are also well suited for searching complex, highly non-linear spaces because they avoid becoming trapped in a local minimum. Their disadvantages are that they are relatively difficult to set up because the user must choose how to represent and encode elements of a solution in the initial population. Additionally, they are difficult to fine tune because they may require modifying the set of decision variables, finding alternate solution encodings, changing crossover and mutation implementations for GA or generating many individuals (particles) and exchanging the information among them for PSO.

If the search problem is continuous and hence deterministic, then it is preferred to use one of the main deterministic methods; these are direct and gradient-based methods. If the function is differentiable then the gradient-based method is the best choice due to its robustness in comparisons with all aforementioned optimization methods. The choice of optimization algorithm is crucial since the results are strongly dependent on the employed algorithm for both accuracy and local minima sensitivity. Evolutionary algorithms are, on the one hand, less sensitive to local minima; on the other hand they are time consuming and constraints have to be included as a penalty term to the objective function. In contrast gradient-based algorithms can lack in global optimality but they allow multiple constraints and are more robust, especially for problems in which a large number of constraints are prescribed [19].

2.6. Characterization of Turbulence

For scientific investigations on turbulence, it is always necessary to determine the turbulence as accurately as possible. For this, a couple of quantities exist which describe the properties of a turbulent flow.

2.6.1. Turbulence Intensity

The strength of the turbulence is called the turbulence intensity. It is defined as the ratio of the root mean square of velocity fluctuation v' to the mean velocity $\bar{v}$ [14]:

$$TI = \frac{v_{rms}}{\bar{v}} \tag{2.79}$$

As it can be seen in Figure 2.9, fluctuation v' can be calculated as:

$$v'(t) = v(t) - \bar{v} \tag{2.80}$$

The root mean square of the velocity fluctuation is given by:

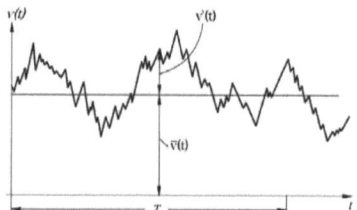

Figure 2.9.: Illustration of the fluctuating velocity [14]

$$v_{rms} = \sqrt{\overline{v'(t)^2}} = \sqrt{\overline{(v(t) - \bar{v})^2}} \tag{2.81}$$

The mean velocity is defined as:

$$\bar{v} = \frac{1}{T} \int_{t}^{t+T} v(t)\,dt \tag{2.82}$$

2.6.2. The Auto-Correlation Function

The turbulent flow is assumed to consists of large eddies with high turbulence energy. This energy is, in turns, delivered to the smaller eddies that have less content of energy, which pass their energy to even smaller eddies. This process continues until all the turbulence energy is converted into heat. So there is a dependence of the smaller eddies

2. Theoretical Background

on the larger ones. The question is how turbulence has changed at time $t+\tau$ compared to the initial situation t. The most common method is the use of the so-called auto-correlation function. Turbulence is a complex, irregular phenomenon, which means that prognoses of the exact behaviour are very difficult. But because of the conspicuous self-similarity, a statistical description is possible. The auto-correlation is defined as [14]:

$$\mathbf{R}(\tau) = \lim_{T \to \infty} \frac{1}{T} \int_0^T v'(t) \cdot v'(t+\tau) \, dt \tag{2.83}$$

Standardization of the auto-correlation function to v'^2 gives the auto-correlation factor ρ:

$$\rho(\tau) = \frac{\mathbf{R}(\tau)}{v'^2} = \frac{1}{v'^2} \cdot \lim_{T \to \infty} \frac{1}{T} \int_0^T v'(t) \cdot v'(t+\tau) \, dt \tag{2.84}$$

A typical distribution of $\rho(\tau)$ is shown in fig. 2.10. At the beginning the auto-correlation is $\rho(\tau = 0) = 1.0$. The correlation disappears with increasing intervals τ until it gets lost. This is where the graph hits the horizontal axis.

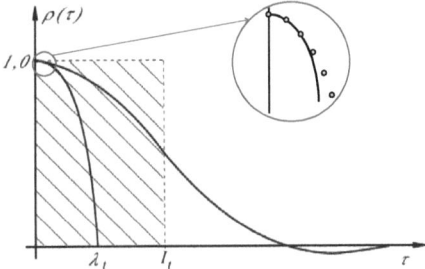

Figure 2.10.: Illustration of the auto-correlation function [14].

2.6.3. Scales of Turbulent Flows

Proceeding from the auto-correlation function, it is possible to determine specific values of the turbulent flow which give information about its structure. Generally, three different scales of turbulent eddies are exist; starting from large to small: Integral scale L, Taylor scale λ and Kolmogorov scale η. They are all connected to each other. The Integral scales (time, velocity and length) can be determined as [14, 11]:

2. Theoretical Background

$$L_t = \int_0^\infty \rho(\tau)\, d\tau \tag{2.85}$$

$$L_v = \overline{v} \tag{2.86}$$

$$L_f = \overline{v}\, L_t \tag{2.87}$$

$$L_g = \frac{1}{2} \cdot L_f \tag{2.88}$$

where L_t represents the time period over which the velocity fluctuations $v'(t)$ are correlated with each other. L_g is the transverse Integral scale that is related with the longitudinal scale L_f.

The micro or Taylor time-scale λ_t can also be calculated from the curvature of the auto-correlation coefficient function at the point $\tau = 0$. It can be found by a curve fitting of the first three points in the auto-correlation curve as shown in figure 2.10 [14, 11].

The turbulence Reynolds number can be calculated as:

$$Re_L = \frac{L_f\, k^2}{\nu} \tag{2.89}$$

where $k = \frac{3}{2}(v_{rms})^2$ and ν is the kinematic viscosity $\nu = 1.45 \cdot 10^{-5}\, m^2 s^{-1}$. The Taylor-scale Reynolds number is connected by:

$$Re_\lambda = \sqrt{\frac{20}{3} \cdot Re_L} \tag{2.90}$$

Now, the Taylor scale parameters can be calculated:

$$\lambda_g = \frac{Re_\lambda\, \nu}{v_{rms}} \tag{2.91}$$

$$\lambda_t = \frac{\lambda_g}{v_{rms}} \tag{2.92}$$

The transverse Taylor micro scales are related to the longitudinal scale λ_f by [11]:

$$\lambda_g = \frac{\lambda_f}{\sqrt{2}} \tag{2.93}$$

The dissipation ε describes how much turbulence energy is converted into heat. It can be determined by:

$$\varepsilon = \frac{15\, \nu\, v_{rms}^2}{\lambda_g^2} \tag{2.94}$$

Knowing that, the Kolmogorov scales can be identified as [11]:

$$\eta = L_f \, Re_L^{-\frac{3}{4}} \tag{2.95}$$

$$\eta_t = \frac{\eta^2}{\nu} \tag{2.96}$$

2.6.4. Spectrum of Turbulence

Turbulence is characterized by different scales of eddies that contain the energy. The energy is distributed between those eddies in such a manner that large eddies contain more energy than small eddies. This energy is delivered to smaller eddies that have less content of energy which, again, is carried to even smaller eddies. The process is named energy cascade, which is completely transfer into dissipation for the very small scales. It continues until all the turbulent energy is converted into heat. Determining the energy for all length scales leads to the energy spectrum. In contrast to the turbulence intensity, the composition of the turbulence does no longer depend on the time t, but it rather changes the fluctuation from time domain to frequency domain. This indicates that the eddies are analyzed by their rotational speed. This can be done with using of the Fourier Transform [37, 11]:

$$a(f) = \int_0^T v'(t) \cdot e^{-i2\pi f t} \, dt \tag{2.97}$$

Thus, it is possible to calculate the one dimensional energy spectrum as:

$$E(f) = \frac{1}{T} |a(f)|^2 \tag{2.98}$$

The spectrum is typically plotted with logarithmic scales for both axes.

3

Numerical Set-up

3.1. Governing Equations

Computational fluid dynamics (CFD) are computer-based simulations for the analysis of fluid flow systems. CFD codes use numerical algorithms to obtain information about the fluid flow. These algorithms are based on the governing equations of fluid dynamics and are applied to a representation of the defined problem [91]. The three governing equations of fluid dynamics are partial differential equations that represent the conservation laws of physics [91]:

- Conservation of fluid mass (Continuity equation)
- Newton's second law (Momentum equation)
- First law of thermodynamics (Energy equation)

The reference quantity are fluid particles. These particles can be thought of as the smallest possible volume part of the fluid, where molecular motions can be ignored. The fluid particles are described with macroscopic properties, such as velocity, pressure, density and temperature [91]. The complete derivations of the governing equations can be found in Durst [35].

3. Numerical Set-up

In a fluid mechanics system the overall fluid mass M stays constant (3.1). Also, the mass of a single fluid particle $\delta m_\Re$, is constant (3.2) [35].

$$\frac{dM}{dt} = 0 \tag{3.1}$$

$$\frac{d\delta m_\Re}{dt} = 0 \tag{3.2}$$

Then, transform it into Eulerian flow specification to obtain (3.3), which is the general form of the continuity equation, where ρ is the density, U_i the velocity in i-direction and x_i the coordinate in i-direction. For the special case of negligible compressibility, the continuity equation is simplified to the incompressible form (3.4) [35].

$$\frac{\partial \rho}{\partial t} + \frac{\partial \rho U_i}{\partial x_i} = 0 \quad (compressible) \tag{3.3}$$

$$\frac{\partial U_i}{\partial x_i} = 0 \quad (incompressible) \tag{3.4}$$

The momentum equation in j-direction is written as (3.5), where τ_{ij} is the molecular momentum transport, p the pressure and g_j the gravitational acceleration.

$$\rho \left(\frac{\partial U_j}{\partial t} + U_i \frac{\partial U_j}{\partial x_1} \right) = -\frac{\partial p}{\partial x_j} - \frac{\partial \tau_{ij}}{\partial x_i} + \rho g_j \tag{3.5}$$

The Navier-Stokes equation is a special form of the momentum equation. For the term τ_{ij} in 3.5 can be shown with symmetry observations that $|\tau_{ij}| = |\tau_{ji}|$. So in the momentum equations, the unknown quantities are, [35]:

$$U_1, U_2, U_3, p, \tau_{11}, \tau_{12}, \tau_{13}, \tau_{22}, \tau_{23}, \tau_{33} \quad 10 \; unknown$$

For the case of an incompressible flow, there are one continuity equation (3.4) and three momentum equations (3.5). Thus, the system of equations thus is not solve-able. The problem can be solved by expressing the six τ_{ij} as functions of $\frac{\partial U_j}{\partial x_i}$ terms. With the term $\tau_{ij} = -\mu \left[\frac{\partial U_j}{\partial x_i} + \frac{\partial U_i}{\partial x_j} \right] + \frac{2}{3}\delta_{ij} \frac{\partial U_k}{\partial x_k}$, the Navier-Stokes equations for a Newtonian fluid with constant viscosity μ becomes:

$$\rho \left[\frac{\partial U_j}{\partial t} + U_i \frac{\partial U_j}{\partial x_i} \right] = -\frac{\partial p}{\partial x_j} + \mu \frac{\partial^2 U_j}{\partial x_i^2} + \rho g_j \tag{3.6}$$

For an incompressible problem, the partial differential equation system has an implicit solution for known boundary and starting conditions. For compressible problems, the energy equation has to be considered additionally, [35]. The mechanical energy equation can be derived from the momentum equations by multiplying them by U_j.

With the introduction of the potential G with the relation $g_j = -\frac{\partial G}{\partial x_j}$ the mechanical energy equation is:

$$\rho \frac{D}{Dt}\left(\frac{1}{2}U_j^2 + G\right) = -\frac{\partial (pU_j)}{\partial x_j} + p\frac{\partial U_j}{\partial x_j} - \frac{\partial (\tau_{ij} U_j)}{\partial x_i} + \tau_{ij}\frac{\partial U_j}{\partial x_i} \quad (3.7)$$

A special form of this equation is the Bernoulli equation for $\tau_{ij} = 0$, $\frac{\partial p}{\partial t} = 0$ and $\rho = const$:

$$\frac{1}{2}U_j^2 + \frac{p}{\rho} + G = const \quad (3.8)$$

The general energy equation is derived from the first law of thermodynamics.

3.2. RANS Equations

A widely applied approximation for turbulent flows in CFD is the approximation by averaging turbulent terms, that only the mean velocity flow field is determined. This leads to the Reynolds Averaged Navier Stokes equations. A turbulent velocity U is split into an averaged mean part and a fluctuating part [50]:

$$U_j = \bar{U}_j + \acute{u}_j \quad (3.9)$$

The term $\bar{U}_j$ represents the time averaged velocity in j-direction and $\acute{u}_j$ the turbulent fluctuating velocity. The continuity equation for the incompressible case becomes:

$$\frac{\partial \bar{U}_i}{\partial x_i} = 0 \quad (3.10)$$

The averaged Navier Stokes equations are [35]:

$$\bar{\rho}\bar{U}_i\frac{\partial \bar{U}_j}{\partial x_i} = -\frac{\partial \bar{p}}{\partial x_j} + \underbrace{\frac{\partial}{\partial x_i}\left[\mu\frac{\partial \bar{U}_i}{\partial x_i} - \overline{\rho\acute{u}_i\acute{u}_j}\right]}_{-(\tau_{ij})_{tot}} + \rho g_j \quad (3.11)$$

By averaging the Navier Stokes equations, an additional term is introduced. This term is a further momentum transport term, due to turbulence. The molecular momentum transport comprises of a laminar and a turbulent part [35]:

$$(\tau_{ij})_{tot} = -\mu\frac{\partial \bar{U}_i}{\partial x_i} + \overline{\rho\acute{u}_i\acute{u}_j} = (\tau_{ij})_{lam} + (\tau_{ij})_{turb} \quad (3.12)$$

3. Numerical Set-up

The term τ_{ij} is called Reynolds stress tensor and is written in full:

$$\overline{u_i u_j} = \begin{pmatrix} \overline{u_1^2} & \overline{u_1 u_2} & \overline{u_1 u_3} \\ \overline{u_2 u_1} & \overline{u_2^2} & \overline{u_2 u_3} \\ \overline{u_3 u_1} & \overline{u_3 u_2} & \overline{u_1^3} \end{pmatrix} \tag{3.13}$$

The Reynolds stress tensor is diagonally symmetrical. It therefore adds six additional unknown variables to the system of equations. The closure problem is still or even increases with the additional non-liner terms of Reynolds stresses. For the determination of the Reynolds stress tensor, turbulence models were used, as will be shown in the next sections.

3.3. Setting up the Problem

3.3.1. Geometry

The blade geometry tested in the numerical investigation work is the wind turbine optimized with TMASO, which is imported as IGES file to Star CCM+. The computational domain was chosen to be cylindrical. The simulations of the three bladed HAWT were performed as steady state simulations, with a geometrically constant inlet wind velocity. Therefore, it was possible to make use of the periodicity of the model. The simulation of a $120°$ fraction of a cylindrical domain with one blade is sufficient. Dimensions are given as multiples of the radius $R = 0.25m$.

For CFD simulations the space for fluids is modelled as a solid body. Solid objects, like the blade, are subtracted from this fluid representation. Here, the blade is enclosed in the inner domain. The inner domain has a radius of 1.04 R and a depth of 0.48 R. The faces for the interfaces to the outer domain and for the periodic interfaces are shown in Figure 3.1 (a). The axis of the cylinder fraction is the axis of rotation. Figure 3.1 (b) shows an overset grid configuration. The blade geometry is enclosed and subtracted from an overset domain. This domain has a depth of 0.2 R and width of 0.44 R. The geometry of the domain is designed to fit in the inner domain. Furthermore, the domain encloses the blade in such a way, that the cell systems with the largest variable gradients, due to the blade fluid interaction, are within this overset domain. On the overset faces, see 3.1 (b), where the gradients are decreased, the overset domain is connected with the inner domain mesh by an overset grid interpolation. The advantage of this configuration is that the overset domain can be moved relative to the inner domain, without changes to the geometry files and mesh. This way, the pitch angle of the blade can easily be varied. On the downside, the overset domain has to be fully enclosed

3. Numerical Set-up

by the inner domain so that the hub has to be neglected in this model. The inner do-

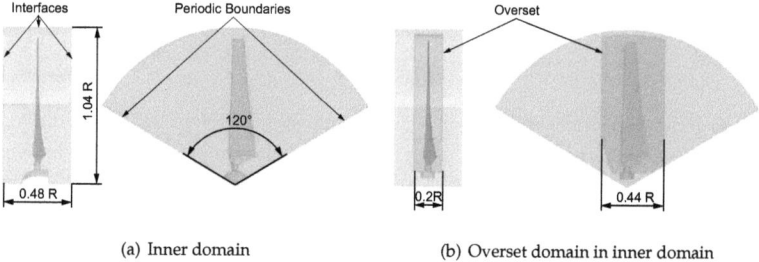

(a) Inner domain   (b) Overset domain in inner domain

Figure 3.1.: Geometrical configuration of inner domain and overset domain

main is enclosed in the outer domain. The connection is at the inner domain interface surfaces that have corresponding interface surfaces in the outer domain. With the periodic boundaries a full cylinder is modelled. The inlet boundary is in a distance of 10 R upstream of the blade. The outlet is 20 R downstream of the turbine blade. The radius of the outer domain is 8 R Figure 3.2. These boundaries are chosen in the farfield, so that the flow over the blade is not affected by interactions with these boundaries. The basic dimensions for the inner domain and outer domain farfield were taken over from Beyer [25]. The geometry files were imported to Star CCM+. The tessellation density

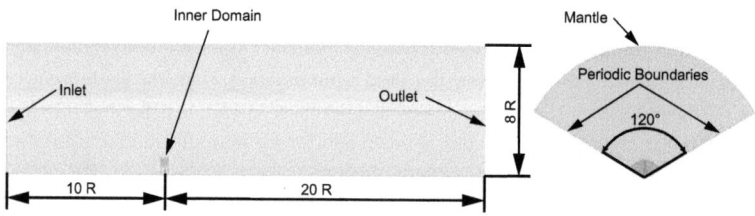

Figure 3.2.: Geometrical configuration of outer domain

defines refinement of surface triangles in the import process. To minimize geometry uncertainties, the tessellation density was chosen very high for the inner overset domain.

3.3.2. Grid generation

The results of a CFD simulation depend strongly on the size and quality of the grid. Therefore, the meshing process is one of the most important steps for a simulation [50].

3. Numerical Set-up

In the current simulations of the current thesis, the grid is generated within the CFD program Star CCM+. The volume mesh consists of polyhedral elements with prisms for the wall regions, where boundary layers exist.

The grid generation is conducted with the use of the options of the $SurfaceRemesher$, to optimizes and improves existing surfaces for volume and prism mashers by re-triangulation, and the $PolyhedralMesher$ that provides an unstructured volume mesh based on the surface mesh, and finally the $PrismLayerMesher$ that adds prism layers to a core volume mesh next to no slip wall surfaces with the defined input parameters [30].

The surface mesh size is set by the base size of the domain. For surfaces of interest, this value can be adjusted. In this thesis, different models and thus meshes are used for the investigations of different aspects. For the wake analysis, a model with hub is used (wake model), that has two domains (outer and inner domain). Based on this model, winglet is tested and compared with the reference rotor (without winglet). Furthermore, a model with a variable pitch angle (pitch model), in which the hub is neglected in order to prevent interactions with the third added overset domain, is conducted.

The polyhedral volume mesh size is defined by the base size. Adjustments are made by volumetric controls that are presented by external CAD volumes for the near and far wakes and for the overlap regions. These are used for controlling the refinement of the entire computational domain. Main mesh settings were defined globally for all domains. The surface growth rate is set to 1.3 and the polyhedra expansion ratio is set to 1.05. CD-adapco [30] suggests values between $1.05 - 1.15$ for incompressible external aerodynamic simulations, for a slow volume growth of the polyhedra. Finally, the prism layer settings, which are dependent on the used turbulence model and the origin of the following values, is set. 21 prism layers are applied with a total thickness of $0.0024m$ and a stretching ratio of 1.2.

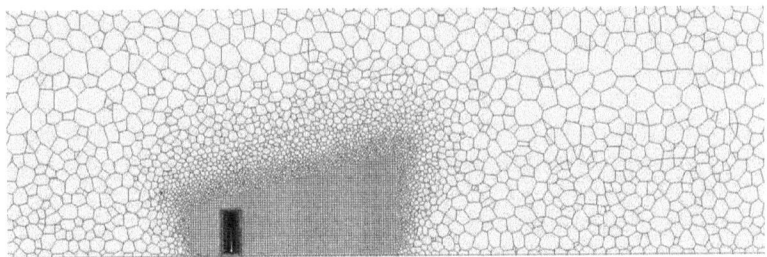

Figure 3.3.: Mesh distribution

The mesh for the pitch angle model is shown in Figure 3.3. The number of cells and

3. Numerical Set-up

Model	Outer domain		Inner domain		Overset domain		Overall	
	Cells $[10^6]$	Faces $[10^6]$	Cells $[10^6]$	Faces $[10^6]$	Cells $[10^6]$	Faces $[10^6]$	Cells $[10^6]$	Faces $[10^6]$
Wake	0.98	6.83	3.03	15.32	-	-	4.00	22.15
Pitch	0.36	2.45	0.27	1.19	4.29	20.40	4.81	24.03

Table 3.1.: Mesh cell and face count

faces for the models is stated in table 3.1.

3.3.3. Physical Model

In the next step appropriate physical models are chosen:

- Space model: For the simulations a three dimensional model is applied

- Time model: The simulations are performed in steady state.

- Material properties: The fluid medium is air, modelled as an ideal gas at a temperature of $T = 15°C$. The density is $\rho = 1.18415 kg/m^3$ and the dynamic viscosity $\nu = 18.5508 \cdot 10^{-6} kg/ms$. The flow can be considered incompressible, as velocities are below $0.3Ma$ [35].

- Flow solver: The fluid is modelled incompressible, so a segregated flow model is applied. The model is based on the SIMPLE algorithm in combination with a Rhie-and-Chow-type pressure velocity coupling. The second order upwind differencing scheme is used. The under-relaxation factors are 0.3 for the pressure and 0.7 for the velocity components.

- Turbulence model: The Reynolds averaged Navier Stokes (RANS) $k - \omega$ SST model is applied. The under-relaxation factor for the $k - \omega$ turbulence is 0.8.

The Menter $k - \omega$ shear stress transport (SST) model was designed for the simulation of external aeronautic flows with adverse pressure gradients and separation. The model combines the advantages of the $k - \epsilon$ turbulence model in non-wall regions with the accurate boundary layer treatment of the Wilcox $k - \omega$ model [68]. It solves two transport equations, one for the turbulent kinetic energy k and one for the turbulence frequency $\omega = \frac{\epsilon}{k}$, where ϵ is the dissipation of turbulence kinetic energy [36].

According to Versteeg [91], the one Spalart Allmaras equation and the two equation models $k - \omega$ and SST $k - \omega$ are suitable for external aerodynamic simulations. The $k - \omega$ SST is suggested for adverse pressure gradient boundary layers and zero pressure gradients. Eleni et. al. [36] ran simulations on a NACA 0012 airfoil with the realizable

3. Numerical Set-up

$k-\epsilon$, Spalart Allmaras and $k-\omega$ SST turbulence model and compared them with experimental data. It is shown that the last model provided the best matching, especially for high angles of attack. In this thesis, off-design configurations with high angles of attack are analysed as well. Therefore, here the Menter $k-\omega$ SST turbulence model is favoured.

3.3.3.1. Near Wall Treatment

An essential part for turbulence models is how the viscous flow in the boundary layers is computed. This determines the requirements for the mesh in the boundary layer region [68]. For the $k-\omega$ SST model in Star CCM+ 8.02 three different wall treatments are available [30]; High y^+, Low y^+ and all y^+ Wall Treatment. For the high y^+ wall treatment ($y^+ > 30$) the shear stresses, turbulent production and dissipation are modelled based on turbulent boundary layer theory. The low y^+ wall treatment ($y^+ \leq 2$) assumes, that the viscous sub layer is resolved so fine, that no wall laws are needed. The all y^+ treatment is a hybrid approach with a blending function for ($2 < y^+ < 30$). Here the all y^+ wall treatment is applied, whereas the low y^+ wall treatment condition shall be kept [30]. The y^+ value is a non-dimensional number characterizing the distance to the wall surface. The number is normed to the wall shear stress in the turbulent boundary layer, which itself is dependent on the Reynolds number. In the following, the distance for the first boundary layer cell from the blade surface is estimated. The Reynolds number (3.14) for the design configuration ($v_1 = 12\frac{m}{s}, \lambda = 4.08$) is approximately the same over the blade span due to the chord distribution ($Re_{c,r/R=0.3} = 57 \cdot 10^3$, $Re_{c,r/R=0.6} = 63 \cdot 10^3$ and $Re_{c,r/R=0.9} = 64 \cdot 10^3$).

$$Re_c = \frac{\rho w c}{\mu} \tag{3.14}$$

The maximum wall shear stress on the blade is given by

$$\tau_w = C_f \frac{1}{2} \rho w^2 \tag{3.15}$$

with the skin friction factor C_f estimated by [81]

$$C_f = [2 log_{10}(Re_c) - 0.65]^{-2.3} \tag{3.16}$$

The wall distance y is defined with the friction velocity u_*

$$y = \frac{y^+ \mu}{\rho u_*} \quad with \quad u_* = \sqrt{\frac{\tau_w}{\rho}} \tag{3.17}$$

3. Numerical Set-up

The higher is the Reynolds number, the lower is the wall distance y. The wall distance is, therefore, defined for the blade segment at 0.9R with $Re_{c,r/R=0.9} = 64 \cdot 10^3$. w for $\lambda = 4.08$ at design wind velocity is $45.67 \frac{m}{s}$. For $y^+ = 2$, the wall distance is $y = 1.2 \cdot 10^{-5} m$. Based on best practice recommendations from CD-adapco [30] for external aerodynamics, 21 boundary layer prisms with a growth rate of 1.2 are applied. With a total prism layer thickness of $2.4 \cdot 10^{-3} m$, the first prism layer has a wall distance of $y = 1.1 \cdot 10^{-5} m$.

3.3.3.2. Moving Reference Frame

In steady state simulations, the movement or rotation of objects can be modelled by additional reference frames. Here, two reference frames are applied. The Outer domain is in the stationary inertial lab reference frame. The Inner domain (and if used the Overset domain) are in a rotational reference frame. The axis of the rotational reference frame is the axis of rotation. The rotational rate ω sets the revolutions per time to be modelled [30]. The steady Navier Stokes equations are altered in the rotating reference frame to [28]:

$$\rho U_i \frac{\partial U_j}{\partial x_i} = -\frac{\partial p}{\partial x_j} + \rho [\underbrace{2\omega \times U_r}_{Coriolis\ acceleration} + \underbrace{\omega \times (\omega \times r)}_{centripetal\ acceleration}] + \mu \frac{\partial^2 U_j}{\partial x_i^2} + \rho g_i \quad (3.18)$$

The centripetal acceleration acts perpendicular away from the axis of rotation, where r is the distance from the axis of rotation, and the Coriolis acceleration perpendicular to the radial velocity component U_r.

3.3.3.3. Overset Mesh

Overset meshes are used to discretize a computational domain with overlapping grids. The main benefits are for the modelling of moving bodies and for parameter studies. Here the pitch angle variation is realized with an overset grid [30]. The configuration consists of two regions. The first region is the overset grid. The outward boundary surfaces of this region are of the type overset mesh. The second domain is the background mesh, which encloses the overset mesh entirely. The grids of the regions then are coupled with the overset interface. Along the overset mesh surfaces an overlapping of both meshes persists. Apart from this overlapping zone, the cells of the background mesh that are within the overset mesh are set to an inactive state. The cells of both domains in the overlapping zone are called acceptor cells. For the acceptor cells, the velocity field is determined by a distance weighted interpolation with cells from both domains [30, 95].

3.3.4. Specification of Boundary Conditions

The above-described models provide a complete and closed system of numerical equations for the given task. With the implementation of boundary condition values, the system can be solved numerically. The values for the wake and winglet models are stated in table 3.2 and for the variable pitch model in table 3.3. The boundary surfaces were introduced with Figures 3.1 and 3.2. The direction of the velocity inlet is normal to the boundary faces. Furthermore, the turbulence intensity is set to a low level at 0.01 and the viscosity ratio to 10.0. For the in-place and periodic interfaces, a tolerance level of 0.05 is applied.

Outer Domain	
Inlet	Velocity inlet
Outlet	Pressure outlet
Mantle	Slip wall
Interfaces	In-place interface
Periodic boundaries	Periodic interface
Inner Domain	
Interfaces	In-place interface
Periodic boundaries	Periodic interface
Blade portions	No-slip wall

Table 3.2.: Boundary conditions for wake and winglet model

Outer Domain	
Inlet	Velocity inlet
Outlet	Pressure outlet
Mantle	Slip wall
Interfaces	In-place interface
Periodic boundaries	Periodic interface
Inner Domain	
Interfaces	In-place interface
Periodic boundaries	Periodic interface
Overset Domain	
Blade portions	No-slip wall
Overset	Overset Mesh

Table 3.3.: Boundary conditions for pitch model

4
Experimental Facilities

4.1. Wind tunnel

The closed-loop wind tunnel of LSTM-Erlangen was used for the experimental investigations Figure 4.1. It is a low-speed tunnel with an open test section. The test section has the following dimensions: width=1.87 m, height=1.4 m and length=2 m. The turbulence intensity is less than 0.1%.

In Figure 4.2, the test section and the employed instrumentations configuration are shown. As can be seen, they were installed on a 3-D traverse system which allowed the probes to be traversed along the test section area. The wind tunnel was equipped with a temperature regulator which kept the flow at a given temperature within 0.1 K. The maximum operating speed at the test section can be set up to 55 m/s in case of open test section. In order to create higher turbulence levels, several grids can be installed at the outlet. For this study, the following two grids were used.

4. Experimental Facilities

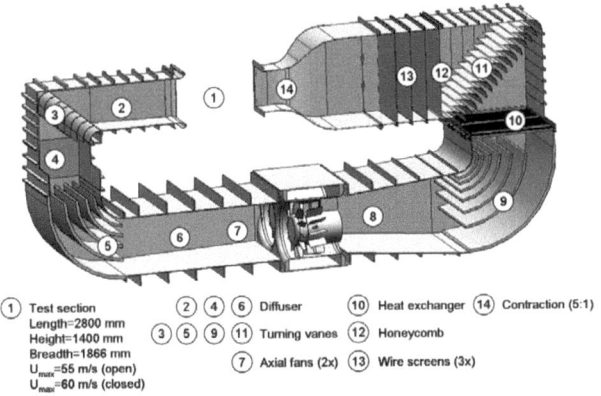

① Test section
Length=2800 mm
Height=1400 mm
Breadth=1866 mm
U_{max}=55 m/s (open)
U_{max}=60 m/s (closed)

② ④ ⑥ Diffuser
③ ⑤ ⑨ ⑪ Turning vanes
⑦ Axial fans (2x)

⑩ Heat exchanger
⑫ Honeycomb
⑬ Wire screens (3x)

⑭ Contraction (5:1)

Figure 4.1.: Wind Tunnel of LSTM-Erlangen

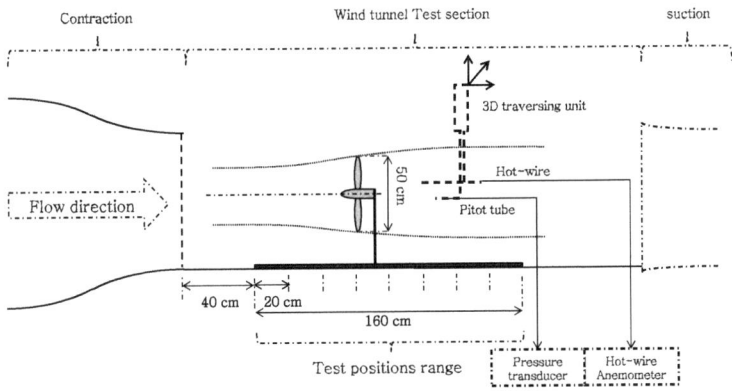

Figure 4.2.: Schematic of the experimental rig in the wind tunnel

4.2. Turbulence Generating Grids

4.2.1. Fine Grid

The fine grid (fig. 4.3) is used to generate medium turbulent flows in the wind tunnel. It is possible to reach turbulence intensities of a maximum of $TI = 0.025$. There are $n_f = 25,620$ quadratic openings with an edge length of $M_f = 8\ mm$ each, so the total open area is:

$$A_{f,\ open,\ total} = n_f \cdot M_f^2 = 1{,}639{,}680 \ mm^2 \tag{4.1}$$

The dimensions of the fine grid are $w_f = 1{,}830 \ mm$ and $h_f = 1{,}400 \ mm$, consequently the opening ratio is:

$$A_{f,\ total} = w_f \cdot h_f = 2{,}562{,}000 \ mm^2 \tag{4.2}$$

$$OR_f = \frac{A_{f,\ total}}{A_{f,\ open,\ total}} = 64\ \% \tag{4.3}$$

4.2.2. Coarse Grid

With the coarse grid installed on the LSTM wind tunnel, which is shown in Figure 4.4, it is possible to generate a high turbulence level with turbulence intensities up to $TI = 0.114$. With a total amount of $n_c = 952$ openings with a squarish size of $A_{c,\ open} = 1{,}600 \ mm^2$ ($M_c = 40mm$) each, it has (in analogy to equation (4.1)) a total opening area of:

Figure 4.3.: The fine grid installed at the wind tunnel outlet

Figure 4.4.: The coarse grid installed at the wind tunnel outlet

$$A_{c,\ open,\ total} = n_c \cdot A_{c,\ open} = 1{,}523{,}200 \ mm^2 \tag{4.4}$$

Since the dimensions of the grid area are $w_c = 1{,}700 \ mm$ in width and $h_c = 1{,}400 \ mm$ in height, the total grid area is given as:

$$A_{c,\ total} = w_c \cdot h_c = 2{,}380{,}000 \ mm^2 \tag{4.5}$$

From here, the opening ratio (OR) can be calculated as:

$$OR_c = \frac{A_{c,\ total}}{A_{c,\ open,\ total}} = 64\ \% \tag{4.6}$$

4. Experimental Facilities

The advantage of the same opening ratio for both grids is that the distance of a inhomogeneous velocity distribution directly at the test section caused by the grids is the same.

4.3. Setup for Performance Measurement

The experimental work was carried out in the wind tunnel of the Institute of Fluid Mechanics (LSTM) at the Friedrich-Alexander-University of Erlangen-Nuremberg (see Figure 4.5). A wind turbine model designed with the TMASO method, as will be explained in chapter 5 (Optimization Procedure), is experimentally tested to verify the analysis method explained in chapter 6 (Performance Analysis).

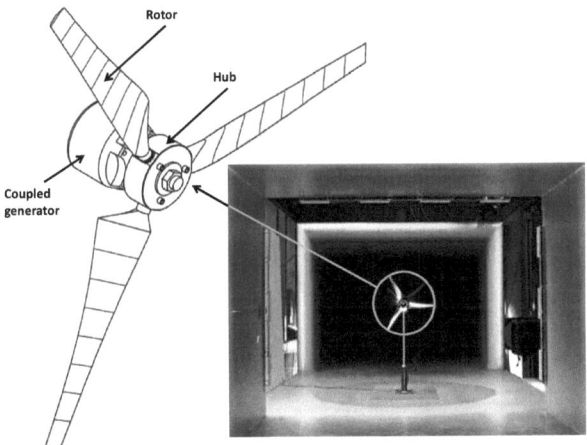

Figure 4.5.: Wind turbine in the wind tunnel

The instrumentation shown in Figure 4.6 was used to measure power extracted from the wind by the turbine and the rotational speed of the turbine. The turbine model was connected to a generator, which is actually a three-phase AC motor (AXi 5330/24). It is given that this motor provides 1 Volt potential per 197 revolutions. This relation was verified with the help of optical measurements and later used to measure the rotational speed by measuring the voltage output of the generator.

The electrical circuit in Figure 4.6 provides a precise control of consumed power by the load (R_{load}), which is equal to the generated total electrical power P_t by the generator. It is designed in such a way that the power consumption of the load can be

4. Experimental Facilities

controlled by adjusting the current (I) passing through it. R_{load} can also be replaced to extend the measurement range or to have the desired operation range.

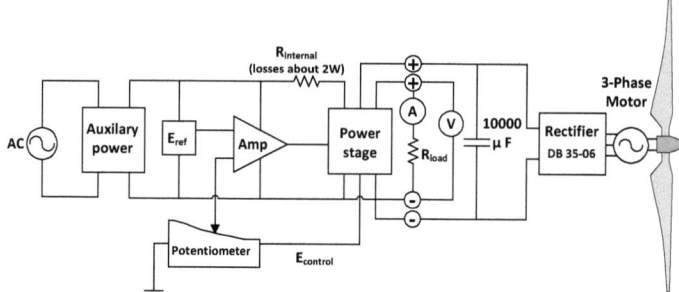

Figure 4.6.: Power measuring circuit

In order to find the mechanical power extracted by the rotors from the wind, it is necessary to know the relation between the measured electrical quantities, torque T_{drive} and rotational speed n of the driving unit (generator). To find out these relations, a set-up as shown in Figure 4.7 was built. The generator is connected to a servo motor, which can keep constant rotational speed. The power measurement circuitry is the same as the one used in wind-tunnel tests, cf. Figure 4.6. An arm is connected at one end to the generator and at the other end placed onto a balance so that torque at the set conditions can be measured.

The measurements deliver the following T_{drive}-n relation at a constant R_{load} and I, which was used in the analysis part:

$$T_{drive} = 7.33 * 10^{-6}\, n + 0.428, \quad [Nm] \tag{4.7}$$

For the chosen values of R_{load}, I and of wind velocity, the generator rotates with a certain rotating speed and produces a potential difference E, which are in turn used to calculate the total electrical power P_t and the shaft power P_s. The total electrical power P_t generated by the generator can be calculated by using the Ohms law, but it will not be used in the present study. Hence, the measurements of E at constant R_{load} and I in the wind-tunnel tests are used to calculate $T(E, I, R_{load})$, $\omega(E)$ and the shaft power P_s:

$$P_s(E, I, R_{load}) = T(E, I, R_{load})\omega(E) \tag{4.8}$$

where,

$$\omega = \frac{2\pi n}{60} \tag{4.9}$$

4. Experimental Facilities

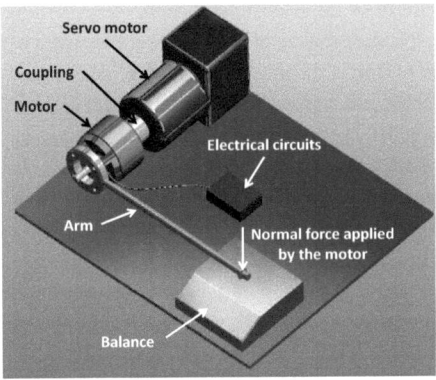

Figure 4.7.: Setup for measuring mechanical power

The power of the wind passing through the turbine blades is:

$$P_w = \frac{1}{2}\rho v_1^3 A \qquad (4.10)$$

Thus, the aerodynamic power coefficient C_p is calculated by:

$$C_p = \frac{P_s}{P_w} \qquad (4.11)$$

In the experiments, C_p values were found out as a function of the tip speed ratio λ by varying the wind-speed.

4.4. Setup for Turbulence measurement

4.4.1. Instrumentations

A Pitot-static tube was fixed on the 3-D traversing system alone for the velocity measurements and beside a hot-wire probe for the calibration of the hot-wire. The velocity was performed by measuring the difference in static and total pressure with a SETRA differential-pressure transducer connected to the Pitot-static tube and by applying the Bernoulli equation. To conduct velocity fluctuating and hence turbulence measurements, a single normal hot-wire connected to an anemometer unit with a constant-temperature bridge CTA was employed. The wire has a length=1 mm and a diameter=5 μm. Velocity calibration of the employed hot-wire was performed at the test section near the exit of the wind tunnel contraction. During the calibration and the measurements, the temperature of the flow was measured with a PT100 temperature sensor in order to correct the measured data for temperature drifts. The hot-wire signals, pressure transducer signal and temperature signal were acquired by a 16-bit A/D converter (NI 6059E DAQ card) installed in a personal computer. The data was recorded for further analyses of spectra and autocorrelation.

The sampling rate (SR) for the measurements was chosen in such a way that a wide scales of fluctuations can be acquired. SR was set to be 20 kHz, with a measuring time of $T = 120\ sec$. Thus, the acquired 2.4 million number of data can provide a wide range of turbulence scales and a certainty of the turbulence intensity (TI) measurement, which is proven experimentally in the present study by increasing the SR until convergence of the TI.

The integral time scales (t_L), and hence the turbulence characteristics scales were calculated via autocorrelation measurements in different stream-wise positions in the test section with the absence of the wind turbine. For that, the acquired data was averaged to 1 second measurement time to exclude the majority of the uncorrelated data.

4.4.2. Investigation Setup

The laboratory scale wind turbine produced by using TMASO optimization method (chapter 5), is used in the present work for the investigations of the turbulence impact on the HAWT performance. The tests are conducted in the closed loop wind tunnel of LSTM. Hence, the rotor blades are specially designed and optimized for this wind tunnel and the generator used.

Recall that the turbulence is generated by two static squared mesh grids; fine and coarse one. Hence, two mainly different turbulence scales are obtained. In addition, the distance between the wind-turbine and the grid is adjusted to have 9 different positions, thus 9 sub-turbulence intensities for each grid, cf. Figure 4.2. As will be shown, the turbulence is nearly isotropic and decays in the flow direction. The developments of Taylor's micro scale (λ_g) and integral scale of the turbulence (L_g) in the flow direction at various Reynolds numbers based the mesh size of the grid (Re_M) are measured. Those measurements are conducted with hot-wire anemometry in the absence of the wind-turbine. The Reynolds number based on the Taylor's micro scale Re_λ is measured with different Re_M and the downstream location. In addition, upstream and downstream turbulence intensities (TI) distributions are measured. Hence, the facility allows to expose the wind-turbine to turbulence with various energy and length scale content.

4.4.3. Calibration of the Hot-Wire

Recall that the hot-wire used in the experiments is connected to the anemometer via a CTA bridge, which keep the wire temperature T_W constant with changing the flow velocity [14]. Following Figure 4.8, a shift in velocity directly leads to the change in voltage that can be recorded. In order to keep the temperature constant, the current needs to be adjusted, again following Ohm's law. This is automatically done by a servo amplifier integrated into the Wheatstone bridge [14].

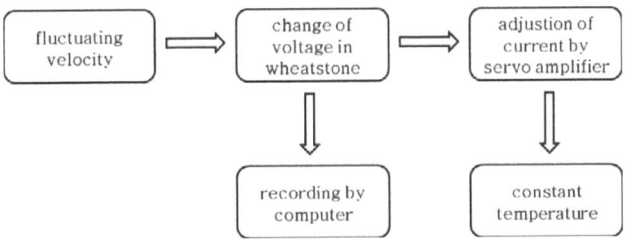

Figure 4.8.: Schematic functionality diagram of a CTA

The calibration is done in two steps. First, the hot-wire probe has to be adjusted. For this step, some characteristics of the wire must be known: The material of the wire and its heat conductivity, the operation temperature T_W of the probe, the electrical resistance of the connection cables and the surrounding temperature T_0 during the calibration. With this information, the hot-wire probe can be adjusted to the anemometer by following the manual instructions. The second step can be done afterwards. The aim is to find the corresponding flow velocity to the output voltage of the amplifier. The relation between the voltage and velocity is non-linear as described by King's law [14]:

$$E_0^2 = A + B \cdot v^n \qquad (4.12)$$

Here, E_0 is the output Voltage, v the flow velocity and n is given as 0.45. In order to find the values for the constants A and B, a minimum two values at known velocities have to be measured. Now, King's law can easily be solved for A and B. The calibration curves are illustrated in Figure 4.9 (a) and in the linearized form 4.9 (b).

E_0 is the voltage corresponding to the calibration temperature T_0. Since the temperature is changing over the day, there will always be a difference in voltage, even for identical velocities. In order to prevent wrong measurement results, a correction of the voltage addicted to the actual surrounding temperature T is necessary. The corrected voltage E as a function of E_0 is given by [58]:

4. Experimental Facilities

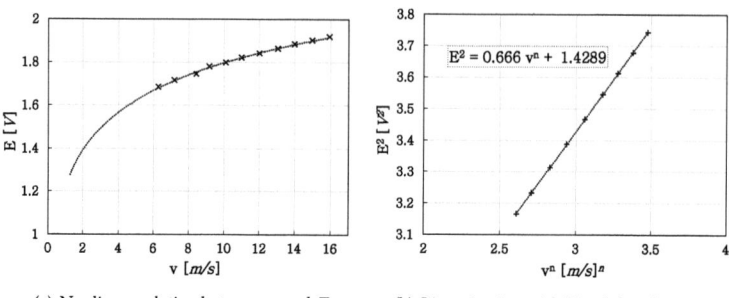

(a) Nonlinear relation between v and E

(b) Linearization with King's law (eq. 4.12)

Figure 4.9.: Calibration curves of the hot-wire

$$E = \sqrt{E_0^2 \cdot \frac{T_W - T_0}{T_W - T}} \tag{4.13}$$

5

Optimization Procedure

In this chapter, two gradient-based optimization algorithm methods to optimize the blade shape and pitch angle of a horizontal axis wind turbine HAWT rotor are presented. Both optimization methods are constrained to the torque-speed characteristic of the coupled generator. The generator characteristic is found by an external experimental set-up. Hence, objectives of the shape and pitch angle optimizations are a maximization of the turbines power coefficient C_p and keep the rated power constant, respectively, while keeping the torque matched to that of the generator.

The aim of the shape optimization is to improve the performance of the selected (stall-regulated) turbine under design and off-design operation conditions, whereas the second optimization suggests an optimum pitch control strategy that helps to convert the turbine control from passive VS-FP to active VS-VP controlled turbine. This serves to enhance its performance at higher wind velocities. In the first section, both methods are applied to a laboratory scale ($0.5\ meter\ diameter$) wind turbine. In the second section, methods are applied to a large scale NREL ($10\ meter\ diameter$) stall-regulated wind turbine.

5.1. Optimization of a laboratory scale WT

In this section, shape and pitch control optimizations methods are applied to a laboratory scale wind turbine (WT) ($0.5\ meter\ diameter$), which is later produced for validations of two methods. The initial design and optimizations formulation procedures are presented in the next subsection.

5.1.1. Aerodynamic Design procedure

The design procedure is based on the combination of Schmitz and BEM theories [67, 47, 42]. The blade profile is designed with Schmitz theory, while the torque and power are calculated by BEM theory. BEM is not selected for the design since it iteratively founds the axial and tangential induction factors a and a' for ring elements along the blade span. The iteration becomes then the bottle neck during optimization. Schmitz theory is favoured for the design over the Betz because it considers the rotation of the wake and hence produces a new profile. In comparison with the design equations of Betz, it is clear that Schmitz method can produce a totally new blade profile owing to including of the rotation effect, where reduction in both axial and relative and increment in the tangential velocities are included (see section Design Theories in chapter 2). Equations are incorporated in a MATLAB code to produce an initial design of the rotor blades. Schmitz can not calculate the induction factors, however, in the present study it will be shown that with some modifications one can calculate them. The necessity of finding the induction factors coming from their need for calculating forces and power, where they contains the information of the magnitude and the direction of the air flow that are needed for relative velocity calculation. Furthermore, with knowing the induction factors a detailed information of the flow expansion, velocity reduction, pressure recovery distance, and the swirl or the downstream weak rotation can be obtained.

In BEM and Schmitz theories a and a' are not constant and dependent on λ_r. This dependency can be obtained by equating two equations of the thrust (axial force), which is obtained from the transport of energy for a ring control volume that moves with the angular velocity of the rotor [67],

$$dF_y = 4a'(1+a')\frac{1}{2}\rho\omega^2 r^2 2\pi r dr, \tag{5.1}$$

and the same force obtained from linear momentum transport, equation 2.46. Equating yields a direct relation of a, a' and λ_r:

$$\frac{a(1-a)}{a'(1+a')} = \lambda_r^2. \tag{5.2}$$

By applying the conservation of the angular momentum, an alternative expression for dP can be obtained from equation 2.47, and write it for λ_r instead of r, to be depen-

5. Optimization Procedure

dent on a, a' and λ_r,

$$P = \frac{1}{2}\rho A v_1^3 \left[\frac{8}{\lambda^2} \int_0^\lambda a'(1-a)\lambda_r^3 \, d\lambda_r \right] \quad (5.3)$$

The maximum power can be achieved when the term $a'(1-a)$ is the highest. An expression for a' can be found from (5.2) and substituting into (5.3), then setting the derivative with respect to a to 0 yields to:

$$\lambda_r^2 = \frac{(1-5a+4a^2)^2}{(1-4a+3a^2)} \quad (5.4)$$

Equating equations (5.4) and (5.2) to obtain a direct relationship between a and a':

$$a' = \frac{1-3a}{4a-1} \quad (5.5)$$

This equation is valid for an ideal HAWT ($C_D = 0$) with wake rotation.

In the original theory of Schmitz, there is no analytical relation for the evaluation of the induction factors. However, Schmitz theory is used here since it can calculate φ as a function of λ_r, and hence with some modifications, a function of the induction factors. From the relative velocity triangles of the Schmitz (Figure 2.5), or writing equation 2.45 in terms of φ_x, and combining with equation 5.5 and solving for the real root gives

$$a = \frac{(5 - \lambda_r \tan \varphi_x) - \sqrt{(\lambda_r \tan \varphi_x - 9)(\lambda_r \tan \varphi_x - 1)}}{8} \quad (5.6)$$

With the final derived equation (5.6), it is possible now to calculate the induction factors from the relative angle φ. Hence, we can proceed with Schmitz. Substituting φ_x into equation (5.6) gives the final distribution of a depending only on λ_r, (2.39). Finally, a and a' becomes dependent solely on λ_r or $\frac{r}{R}$. A comparison with BEM induction factors showed that they are almost identical for radius ratio higher than 0.2, as expected for the same assumptions, cf. Figure 5.1. Normally, the blade of the HAWT is cut at 0.15 to 0.25 radius because the torque produced within this region is negligible, and because it adds additional unwanted blockage that distributes the oncoming flow and consequently increases the axial load.

After finding the induction factors for the designed blade, the torque T, the axial force F and the power P can be calculated by using equations 2.57 and 2.58. Or by using the blade element instead as:

$$T_{rotor} = B \int_0^R dT = B \int_0^R r \, dF_x = B \int_0^R r \frac{1}{2} \rho w^2 c C_x dr, \quad (5.7)$$

5. Optimization Procedure

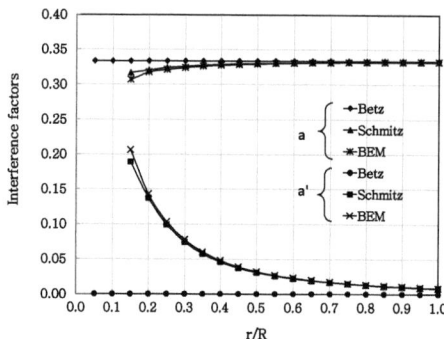

Figure 5.1.: Distribution of a and a' with the rotor radius ratio

where B is the number of blades and $C_x = C_l \sin(\varphi)$ with zero drag coefficient. The relative velocity w can be found either by BEM as a function of a and a' or by Schmitz as a function of φ [47].

In order to include the profile loss (drag loss) $\eta_{profile}$, equation 2.65, should be multiplied by the local torque (dT), on a differential ring element along the span which is calculated in the absence of profile loss. The effect of the tip loss, which is produced from the pressure difference between the low upper and high lower pressure at the rotor blade tip, can be included by multiplying η_{tip}, equation 2.64, with the T_{rotor}.

This formulation is applied to design the blade geometry under given wind and rotational speed to calculate the generated torque by the rotor.

5.1.2. Formulation of the shape optimization

In reality, when a freely rotating wind turbine is coupled to a drive unit, it will adjust itself to the rotational speed characteristic of the drive unit and find the proper rotation speed for each wind velocity. The drive unit can be either a direct one or it comprises a gearbox and a generator. An optimization method is then needed to change the blade shape in order to capture the maximum power from the wind under the torque rotational speed constraint of the drive unit. In the present work, the developed method is called as Torque-Matched Aerodynamic Shape Optimization (TMASO). The objective is to find the maximum power coefficient, which is:

$$C_p = \frac{\omega T_{rotor}}{\frac{1}{2}\rho v^3 \pi r^2} \tag{5.8}$$

5. Optimization Procedure

Hence, the objective function is formulated as:

$$min\,(-C_p) \tag{5.9}$$

As the rotor coupled to the drive, T_{rotor} should match the torque of the drive unit $T_{drive}(n)$. The difference between them should be minimized so that torque matching can be achieved. For this purpose, the following constraint is used during the optimization.

$$ceq(x) = \sqrt{(T_{rotor} - T_{drive})^2} = 0 \tag{5.10}$$

Note that $T_{drive}(n)$ which should be known for an existing wind turbine or it is measured separately before proceeding with optimization. In the present study, it is measured with an experimental set-up as will be shown in the next chapter, see Al-Abadi [73].

The procedure optimizes the wind speed and rotational speed together with the independent decision parameters of the rotor blade shape. The shape of the rotor blade is defined by the local pitch angle $\beta(r)$ and chord length $c(r)$ along the blade and the sectional profile as well [17, 67]. In the present study the sectional profile is kept the same, hence, it is excluded from the decision parameters. The optimization is done by using the gradient based function $fmincon$ with the active-set algorithm in MATLAB. As explained before, $fmincon$ can search for the minimum of a constrained non-linear multi-variable function. Variables are wind velocity v_1 and the angular speed ω that are formulated as:

$$\begin{aligned} x(1) &= v_1 \\ x(2) &= \omega \end{aligned} \tag{5.11}$$

In addition to the constraint, limitations and bounds are needed before moving on with the procedure. These are the rotor radius R, which should not exceed one tenth of the wind tunnel test section area to avoid any wall-model interaction effects. Thus, it is selected to be 0.25 m. The bounds are the wind velocity v_1 and the rotor's angular speed ω, since the wind tunnel velocity is limited as well as the motor rotational speed. The lower bounds of v_1 and ω are set to be as,

$$\begin{aligned} 8 &\leq v_1 \leq 16 \\ 155 &\leq \omega \leq 275 \end{aligned} \tag{5.12}$$

Large scales wind turbines generally operate under high Reynolds numbers. When down-scaling, wind turbines expose low Reynolds numbers, thus, it is important to select an airfoil profile that perform better under that number. Hence, a high lift low

5. Optimization Procedure

Reynolds number profile of SG6043 is selected [45]. It is kept along whole span. The aerodynamics characteristics of the SG6043 airfoil at the design angle of attack $\alpha_d = 8°$ are analysed with XFOIL code in order to highlight its advantage over the large-scale wind turbine common airfoil NREL S809 [88, 85] when operating at low Reynolds number of 65000, which is the operating Reynold's number of the model in the wind tunnel, Figure 5.2. It is clear from the Figure that at a low value of Reynolds number the SG6043 can perform better and gives $C_L = 1.4$ and $C_D = 0.0286$, thus $GR = 49.2$ owing to the thin boundary layer and the effective stall delay which led to a slight drop in pressure, Figure 5.2(a), whereas the S809 gives $C_L = 0.193$ and $C_D = 0.081$ so $GR = 2.38$ due to stall, which causes a full detachment of the boundary layer, thus a dramatic drop in pressure, Figure 5.2(b).

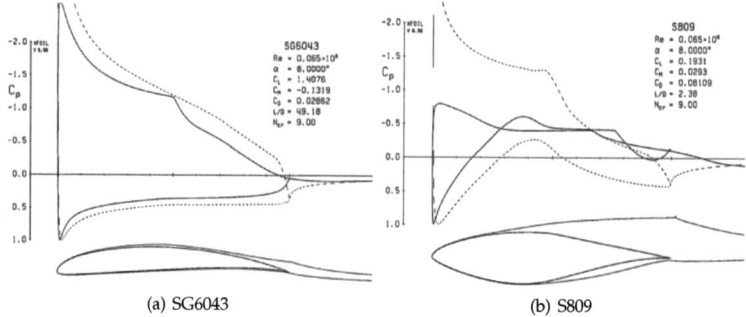

(a) SG6043   (b) S809

Figure 5.2.: Aerodynamic characteristics at low Reynold's number of 65000 and $\alpha = 8°$

The optimization procedure is started with an initial blade design by applying Schmitz method at $\lambda = 7$, which is the typical value for most of three blades wind turbines. In order to highlight the influence of the losses types, two different torque matched blade shape optimizations were performed. First, optimization was conducted without considering the losses. Second, profile and tip losses were included in the optimization. The optimizations ended with chord and pitch angle distributions along the blade span as shown in Figure 5.3(a). The results show that the blade has to cover the power reduction caused by the losses with additional increase of both c and β.

The resultant optimized blade shape that includes both profile and tip losses is considered to be the final optimum design as shown in Figure 5.3(b). The optimized rotor operates at $v_1 \approx 12\ m/s$ and rotates with $\omega \approx 196\ rad/s$. Hence, it delivers a maximum $C_p \approx 0.43$ at $\lambda \approx 4.08$ and Reynolds number of about 65000. During the optimization cycles and for the chord, torque, power and efficiency calculations, it is mandatory to calculate the airfoil characteristics C_L and C_D for every combination of α

5. Optimization Procedure

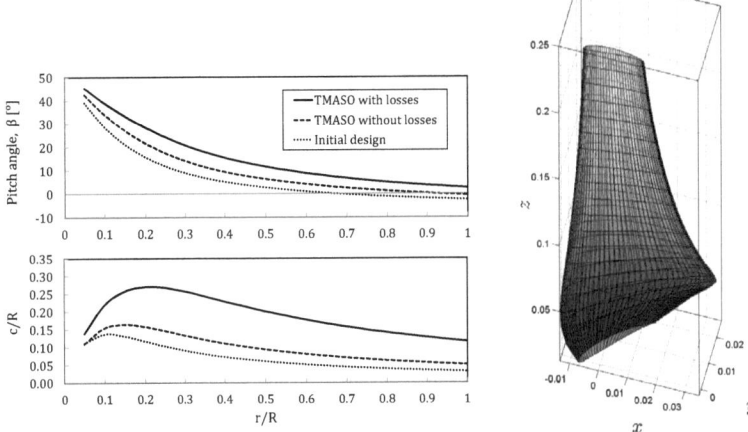

(a) Initial and optimized designs of pitch angle and chord distribution along the rotor radius ratio

(b) Torque-Matched optimum blade final shape

Figure 5.3.: TMASO optimized wind turbine

and Reynolds number simultaneously. Therefore, the TMASO program is successfully interfaced with XFOIL. Figure 5.4 illustrates the optimization procedure diagram, in which the inputs are R, B, ρ, α and the airfoil type. The initial rotor design will be later performed by Schmitz method and gives c_i, β_i and λ_i. In parallel, input data inters XFOIL and returns C_L and C_D that were used in the initial design calculation. Once we obtain the initial design, it is set in a decision optimization loop and formulated as a function of decision variables c and β that should work together with the free variables v_1 and ω to satisfy the objective function (5.9) while keeping the constraint (5.10). The constraint is evaluated for every iteration by calculating T_{rotor} of the current iterated rotor shape with BEM method and involving the $T_{drive}(n)$ experimental equation (equation 4.7 in section 4.3). The iteration continue until it converges with a maximum C_p value and gives the final shape c, β, operational conditions v_1, ω, λ, Re and performance characteristics P, T, C_p.

The torque matched optimized blade is considered as a point design, which gives the maximum power at one wind velocity with its associated rotational speed of the rotor. During the optimization, the relative angle is at its optimum value of φ_x.

5. Optimization Procedure

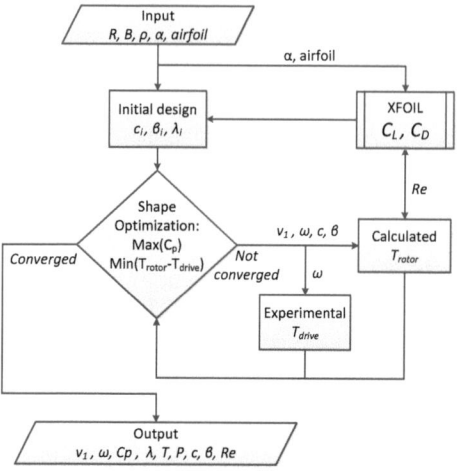

Figure 5.4.: Torque-Matched aerodynamics shape optimization procedure diagram

5.1.3. Formulation of the pitch-control optimization

An approach of the pitch optimization procedure, which is called as Torque-Matched Pitch Control Optimization (TMPCO) is produced and applied to the laboratory scale wind turbine. Its objective is to determine a pitch control angle $\beta_{c(v)}$, which keeps the torque and power constant at wind velocities higher than the rated velocity $v_1 > v_N$. In Figure 5.5, the aerodynamic forces at the turbine of a pitch regulation are illustrated. As the wind velocity v increases with constant tip speed u, w will increase. Thus, α increases until the stall. In order to keep the rating power constant, since the rotational speed ω is constant, the torque should be kept constant as well. It is done by keeping the tangential force ($F_x = F_L \sin\varphi - F_D \cos\varphi$) constant by increasing β with pitch control angle β_c to reduce the angle of attack from α to α' and hence the aerodynamic forces.

In this procedure, the wind velocity increases from $13m/s$, where rating power of $P_N = 100W$ is reached, to $16m/s$. Due to rating conditions, the rotational speed remains constant at $1872rpm$ and hence angular speed is $196rad/s$. Chord and twist distributions are taken from the optimized design shape, Figure 5.3. The mean value of angle of attack α_m is considered during optimization as a representable for α distribution along the radius for simplification, with excluding the extreme values at tip and hub regions for accuracy. The only variable that needs to be adjusted is the pitch control angle $\beta_{c(v)}$. It is calculated by the function $fmincon$ for a range of wind velocities.

5. Optimization Procedure

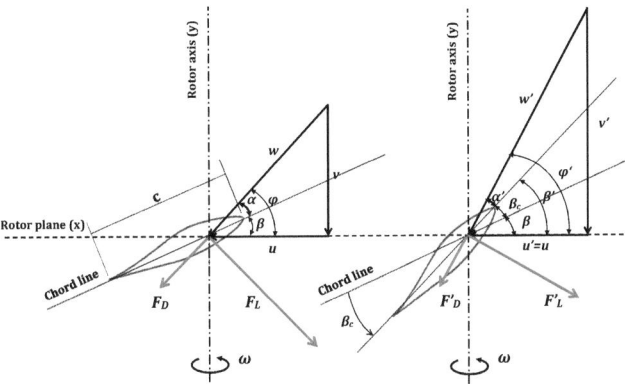

Figure 5.5.: Insertion of β_c to reduce α and thus tangential force and torque during increasing wind velocity.

Its bounds are limited to

$$0° \leq \beta_{c(v)} \leq 30° \qquad (5.13)$$

The objective function is formulated in such a way that the calculated power should not differ from the rated one since the aim is to keep the rotor power constant at $P_N = 100W$ for the wind velocities higher than the rated velocity ($v_1 > v_N$), for the laboratory scale turbine of the design velocity $v_1 = 12m/s$, the rated velocity is considered to be $v_N = 13m/s$. Hence, the objective function is formulated as:

$$min \left(\sqrt{(P_{rotor} - P_N)^2} \right) \qquad (5.14)$$

The power is calculated by integrating dP of equation 2.37 and by considering tip and profile losses. The constraint is formulated where the rotor torque should still match the rated torque of the drive unit

$$ceq(x) = \sqrt{(T_{rotor} - T_N)^2} = 0 \qquad (5.15)$$

T_{rotor} is calculated from equation 5.7, whereas T_N should be known for an existing wind turbine or in this case it is calculated from the known rated power and angular speed as $T_N = P_N/\omega = 0.51 Nm$. Under these conditions, a pitch control angle distribution with wind velocity is obtained, as illustrated in Figure 5.6. The curve fitting equation of the pitch control angle distribution of TMPCO, which is needed in the validation

5. Optimization Procedure

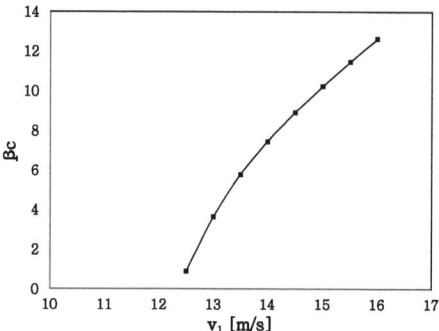

Figure 5.6.: Pitch control angle distribution of TMPCO over wind speed for the laboratory scale wind turbine.

section (chapter 6) is:

$$\beta_{c(v_1)}[°] = -0.0377v_1^4 + 2.2959v_1^3 - 52.57v_1^2 + 538.93v_1 - 2085.3 \quad (5.16)$$

Figure 5.7 illustrates the optimization procedure diagram of TMPCO, in which the input data are R, B, ρ, α T_N, P_N, ω_o, v_N and the airfoil type.

The initial rotor design will later either be performed by using the Schmitz method or by insertion of an existing shape to obtain c_i, β_i and λ_i. Once we obtain the initial design, it is set in a decision optimization over v_1 loop and formulated as a function of the decision variable β_c, which correct the β_i to be a β' that satisfy the objective function $min(P_{rotor} - P_N)$ at the design angular speed ω_o while keeping the constraint $min(T_{rotor} - T_N)$. The constraint is evaluated for every iteration by calculating T_{rotor} of the current iterated rotor shape with the BEM method and involving the T_N. The iteration continue until it converges with a minimum difference in power and gives the final distribution of β_c as a function of wind velocity v_1.

5. Optimization Procedure

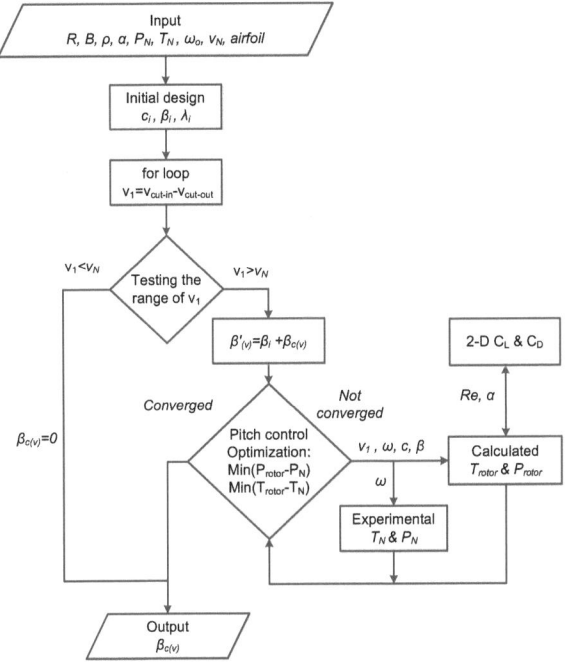

Figure 5.7.: TMPCO procedure diagram.

5.2. Optimization of NREL stall-regulated WT

In this section specifications of the NREL UAE phase-VI 10kW rotor explained in A.1 are conducted as input parameters for the procedures of applying both TMASO and TMPCO methods on a real scale wind turbine.

5.2.1. Settings of Input Parameters

For the optimization and analysis proceedings of the NREL UAE phase-VI rotor, a number of sections is set to 10 at blade's spanwise locations. Air density ρ is set to $1.16 kg/m^3$. Chord length $c_{(r)}$ and pitch angle $\beta_{(r)}$ distributions are set by curve fitting equations for the curves in Figure A.2, as:

$$c_{(r)}[m] = -0.1011r + 0.8639 \qquad (5.17)$$

5. Optimization Procedure

and,

$$\beta_{(r)}[°] = -0.0825r^5 + 1.57r^4 - 12.2r^3 + 48.7r^2 - 102.2r + 92.1 + 3 \quad (5.18)$$

where tip pitch angle of 3 is added to the end of the equation to make it more flexible for different tip pitch angles investigations. A curve fitting equation for the torque before and after the rated torque is reached, Figure A.6 is calculated as:

$$T_{(n)}[Nm] = 0.00000531\, n^4 - 0.001196\, n^3 + 0.10069\, n^2 - 3.731\, n + 51.45, \; for\; n < 71$$
$$T_{(n)}[Nm] = 18.77n, \; for\; n \geq 71$$
$$(5.19)$$

The rotational speed is kept constant at $71.6rpm$. The number of blades, namely 2, the sectional airfoil $S809$ and the blade radius of 5 are kept the same. One initial angular speed, one initial wind velocity and design angle of attack have to be known as inputs for the TMASO. As a result of a constant rotational speed at $71.6rpm$, the initial angular speed is $7.5rad/s$ ($\omega = 2\pi n/60$). Initial designed wind velocity is calculated by following equation:

$$v_{1,o} = \frac{\omega_o R}{\lambda_o} \quad (5.20)$$

The value of λ_o is required, which is taken as $\lambda_o = 5.2$. It is shown in Figure A.5 that the maximum power coefficient C_{Pmax} of the NREL UAE phase-VI rotor at $\beta_o = 3°$ and the design wind velocity $v_{1,o} = 7.2m/s$. Modification for the design angle of attack (α_d) is suggested; to do that, a range of angles within a range of $3°$ to $9°$ were tested. C_{Pmax} and maximum glide ratio GR_{max} are found at angle of attack of $6°$, as illustrated in Figures 5.8 and 5.9.

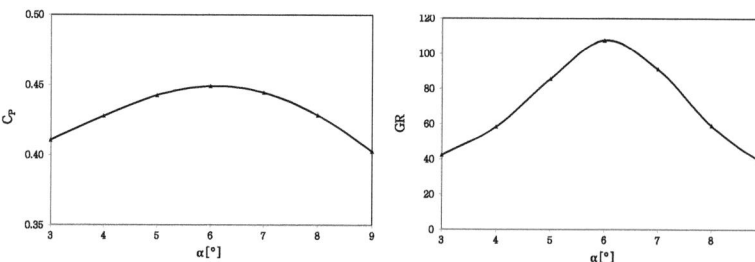

Figure 5.8.: Power coefficient over angle of attack calculated by TMASO.

Figure 5.9.: Glide ratio over angle of attack calculated by TMASO.

5.2.2. Shape optimization of 10kW stall-regulated HAWT

The torque matched aerodynamic shape optimization approach (TMASO) is used. The same approach of changing the rotor shape is kept to reach maximum power coefficient while keeping the torque matching with the drive unit.

The initial design of the blade, which includes chord length and twist angle from equations 5.17 and 5.18, is used as input. Airfoil characteristics C_L, C_D and GR for each iteration of α and Reynolds number are found simultaneously by XFOIL. The objective is to find the maximum power coefficient, thus it is formulated as in equation 5.9. The constraint is still the same of matching the drive torque $T_{drive}(n)$ as in equation 5.10. $T_{drive}(n)$ is calculated by equation 5.19 before and during rated operating. The upper and lower bounds of ω are limited to ensure keeping the optimized rotational speed in the range of the initial one (generally wind turbine generators are set to operate at a defined rotational speed, which represent the capability limit that cannot be exceeded). Bounds of the velocity are more stretched.

A new velocity is found, which indicates a new operation condition at new value of λ and hence new values of P_w, P and T_{rotor}, which should be equal to T_{drive}. The optimized rotor shape parameters $\beta_{(r)}$ and $c_{(r)}$ are calculated by using Schmitz equations 2.40 and 2.43. Tangential and axial induction factors are calculated by using equations 5.6 and 5.5 To consider tip and profile losses along the blade, they are included in the calculation of maximum power coefficient by equation 2.64 and 2.65.

The achievement of TMASO is an increment of the power coefficient, torque and power. The optimized rotational speed is kept exactly the same as the initial one, whereas the optimized tip speed ratio increases from 5.2 to 5.52 and wind velocity decreases from 7.2 to $6.8 m/s$ at the design angle of attack of $6°$. A special treatment is considered for the chord length, where the improvement in performance tends always toward increasing chore values. Hence, the maximum possible of the chord is always expected to be provided by the optimization procedure. Therefore, an upper limit for the chord is done to have a realistic optimal design. Initial and optimized results are listed in table 5.1. Chord length and twist angle along the blade are shown in Figure 5.10. In this figure, it is well cognizable that the chord length distribution is considerably increased along the whole blade by TMASO in contrast to the initial design. The twist distribution reaches lower values until $r/R = 0.3$. After this point the trend of the optimized twist distribution is increased in comparison to the initial design. Curve fitting equations of the optimized chord length and twist angle distributions over r/R are found. These equations, which were calculated for maximum power coefficient at the design point, are used later as input profile in the Performance prediction of the stall-regulated shape optimized WT section.

5. Optimization Procedure

	initial values	optimized values
$v_1[m/s]$	7.2	6.8
$\omega[rad/s]$	7.5	7.5
λ	5.2	5.52
$n[rpm]$	71.62	71.65
$T[Nm]$	757	869
$P[W]$	5677	6520
C_p	0.334	0.456

Table 5.1.: Comparison between initial and optimized values by TMASO.

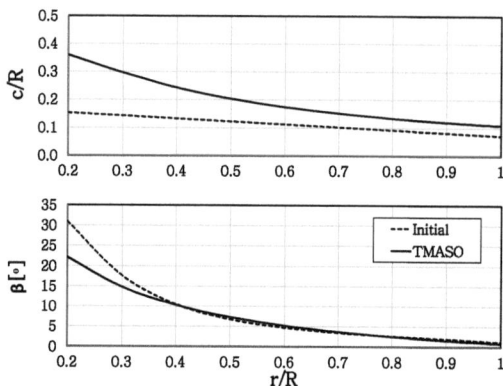

Figure 5.10.: Initial and optimized chord length and twist angle distributions along the blade by TMASO.

5.2.3. Pitch control optimization of stall-regulated WT

The same approach of TMPCO is applied for a large scale stall-regulated wind turbine of NREL phase-VI to verify the applicability of the method on different scales by changing the control strategy to a variable-pitch one. Following TMPCO procedure with considering the NREL phase-VI turbine specifications, where the wind velocity increases from $10m/s$, at which rating power is reached, to $25m/s$. Due to rating conditions, the rotational speed remains constant at $71.6rpm$ and hence angular speed is $7.5rad/s$. Chord and twist distributions are taken from the initial design equations (5.17) and (5.18). The mean value of angle of attack α_m is considered during optimization as a representable for α distribution along the radius for simplification. Lift and drag coefficients are calculated by curve fitting equations for the sectional airfoil of S809.

The rated power is $P_N = 10kW$ for wind velocities higher than the rated velocity

5. Optimization Procedure

($v_1 > v_N$), Figure A.6, whereas the rotor torque should still match the rated torque of the drive unit, T_N which is calculated as mentioned in section Shape optimization of 10kW stall-regulated HAWT by equation 5.19 before and during rated rotational speed.

Under these conditions, a pitch control angle distribution with wind velocity is obtained, as illustrated in Figure 5.11. Pitch control angle distribution of NREL $5MW$ turbine is plotted in this figure as well for a comparison reason. Curves show nearly similar trends with an upper shifted of the $10kW$ values of $\beta_{c(v)}$, since the NREL $10kW$ has a rated velocity $v_{N(10kW)} < v_{N(5MW)}$. Thus, β_c of the $10kW$ starts earlier and continue until the maximum velocity of ($v_{1max} = 25$). The curve fitting equation of the pitch control angle distribution of TMPCO, which is needed in the validation section Performance Analysis of the Pitch-Control Optimized HAWT, is:

$$\beta_{c(v_1)}[°] = -2.98*10^{-5}v_1^6+3.37*10^{-3}v_1^5-0.158v_1^4+3.90v_1^3-53.9v_1^2+394.9v_1-1193.9 \quad (5.21)$$

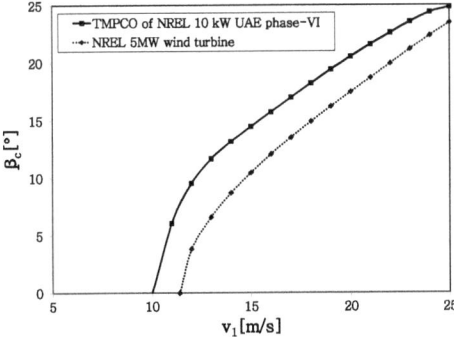

Figure 5.11.: Pitch control angle distribution of TMPCO of $10kW$ and NREL $5MW$ wind turbine [55].

6
Performance Analysis

In this chapter, a developed performance prediction analysis method called torque-matched aerodynamics performance analysis (TMAPAM) is presented [20]. It is first applied to predict the performance of the laboratory scale wind turbine. The method is later adjusted to predict the performance of the NREL phase-VI $10kW$ stall-regulated turbine for both shape and pitch control optimizations. Additionally and to verify the applicability of the pitch control performance prediction method for a large scale wind turbine, it is applied on the NREL $5MW$ wind turbine.

6.1. Performance Analysis of Laboratory Scale WT

The analysis method is developed in order to be applied to any existing rotor. The laboratory scale optimized wind turbine, which is produced by TMASO as presented in section 5.1.2, is first selected to predict its performance. The same rotor is reproduced by using Schmitz design equations 2.40 and 2.43, as shown in Figure 5.3. The same design parameters are set to have the same design conditions and profile of the reproduced wind turbine. Then, the shape is fixed and the off-design condition were set to predict the performance.

The BEM method is used to predict forces torque and power. While operating, the wind turbine adjusts itself to the rotational speed-torque characteristic of the drive unit and finds the proper rotational speed for the present wind velocity and the settings

6. Performance Analysis

of the drive unit. A developed optimization method is introduced into the analysis so that the method can be used to test the turbine rotor over different oncoming wind velocities. However, transient variations of wind speed and transient response of wind turbine to those are not considered. As the rotor coupled to the drive, $T_{rotor}(n)$ must be equal to the torque of the drive unit $T_{drive}(n)$ at any rotational speed. Hence, in the analysis the difference between them should be minimized so that torque matching can be achieved. For this purpose, the same constraint used in the shape optimization is reused as an objective function in the optimization part of the analysis:

$$min|T_{rotor} - T_{drive}| \qquad (6.1)$$

Note that, $T_{drive}(n)$ is measured separately in this study as shown in section 4.3.

In order to monitor the rotor performance at different oncoming wind velocities, the torque matched C_p calculations were performed by using a nested optimization procedure shown in Figure 6.1. As the rotor will be subjected to wind velocities deviating from the optimum one, the relative angle φ has to be calculated at each ring element along the span of the turbine. Equation (2.37) is used for this purpose, since it represents the power which can be extracted by the rotor at any velocity (deviated from the design velocity). The tangential force on the same element and, subsequently, the power extracted can be calculated by using the airfoil theory equation (2.49) with writing w in terms of the known w_1, which can be found from equation (2.33), to get:

$$dP = B\frac{1}{2}\rho(w_1 cos(\varphi_1 - \varphi))^2 cdrC_L sin(\varphi)r\omega \qquad (6.2)$$

Equating equations (2.37) and (6.2) gives:

$$\frac{tan(\varphi_1 - \varphi)sin(\varphi)}{C_L} - \frac{Bc}{8\pi r} = 0 \qquad (6.3)$$

The first term of this equation is an unknown function of r and φ, whereas the second term is a known function of r. Thus, for each ring element at radius r with a spanwise step of dr, the equation will be function of φ only. Hence, φ which is the real root of (6.3) is found by using the $fzero$ optimization function of MATLAB. $fzero$ is a scalar nonlinear zero finding function for solving a single nonlinear algebraic equation which finds the root of a continuous function of one variable near a selected initial value of x_0. Its algorithm involves a combination of bisection, secant, and inverse quadratic interpolation methods [5]. Since C_L is a function of α, so it is written in terms of $(\varphi - \beta)$. The $fzero$ function needs an initial guess for φ, for which an expression for adjusting φ_x (2.39) is used.

The external optimization loop considers the torque matching. For each ω, it sends

6. Performance Analysis

v_1 to $fzero$ function, so that the new $\varphi(r)$ distribution and, consequently, T_{rotor} is calculated. He checks whether torque matching of equation (6.1) is achieved and decides either to continue optimization iteration with a new v_1 or to stop. Finally, the program returns a set of v_1 with corresponding ω, for which torque matching is satisfied. Using this data, besides $\lambda = \omega R/v_1$, the power coefficient is calculated by using equation (5.8). Hence, the objective function is minimizing the Torque difference and calculating C_p, which should be equal to C_{Pmax} only at the design λ. In order to account deviations in Re and α, which is expected since the operational conditions are deviated from the designed one, interpolation equations are used which are based on the experimental airfoil data of the airfoil SG6043. This analysis method is called torque-matched aerodynamics performance analysis method (TMAPAM), its procedure is illustrated in Figure 6.1, [20].

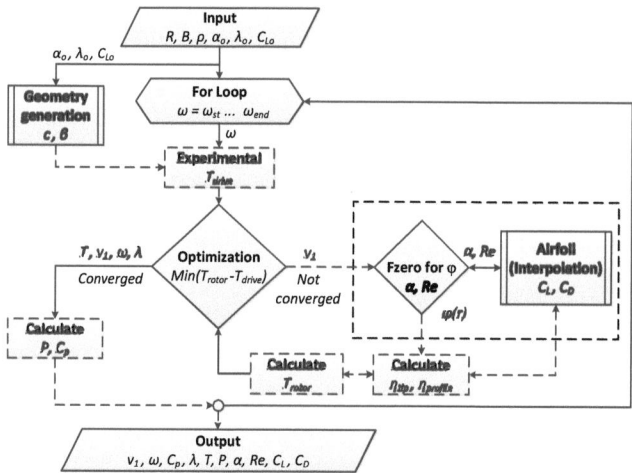

Figure 6.1.: Analysis flow diagram

In the analysis, in order to avoid the unstable stall region we kept the upper range of angles under any value such as the stall value with the following inequality constraint:

$$(\alpha - \alpha_{stall}) \leq 0. \tag{6.4}$$

Furthermore, to incorporate the profile and tip losses, the efficiency relations (2.64) and (2.65) are used.

Results from the experiments and the torque matched performance analysis method TMAPAM are compared in Figure 6.2. The analysis includes profile and tip losses.

6. Performance Analysis

There are three important regions in this figure. The first one is the design region (or maximum power region), where α is near to its design value. The comparison shows a good agreement with the experimental results in this region. The second region is the stall region where the λ gets its lowest values, which means that α will reach α_{stall} that, in turns, causes a dramatic drop in the aerodynamic forces, consequently, a drop of C_P. Unfortunately, it was not possible to measure many points within the low λ region since the stall dominates all or part of the rotor area. The third region is the low α region. The power drops in this region with increasing λ due to the drop of C_L associated with the drop of α.

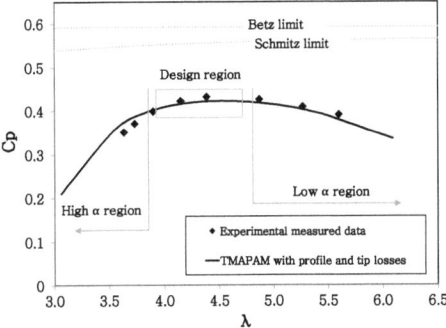

Figure 6.2.: Comparison between the calculated and the measured power coefficients

Figure 6.3 shows that torques are identically matching and following the experimental T-n liner equation (4.7) except at low rotational speed range of $n < 1620\ rpm$ which indicate that the stall is started to effect here. The stall leads to detaching the boundary layer, thus producing an uncertainty of the 2-D data of XFOIL used for predicting the aerodynamics characteristics. Therefore only few points are obtained from TMAPAM at the near stall region.

The torque matched relation between v_1 and ω is shown in Figure 6.4. It is almost linear for high ω range $180 - 275 rad/s$. For low values of $\omega < 180$, since $T_{drive}(n)$ relation of equation (4.7) is linear whereas $T_{rotor}(n)$ decreases due to stall. Thus TMAPAM suggested an increasing in the v_1 to compensate that reduction to keep the rotor rotating while torques are matching. This is done until $\omega = 170 rad/s$, which is associated with $n = 1620 rpm$ and $\lambda \approx 3.5$ that explains the torque perfect matching until this point in Figure 6.3 and the dramatic drop of C_p in Figure 6.2.

Figure 6.5, reveals the off-design α distribution along the radius ratio (r/R) for different λ's. At low values of λ the stall dominate a wide range of the entire blade starting from inboard region. This local stall starts to subsiding towards the hub as λ increases

6. Performance Analysis

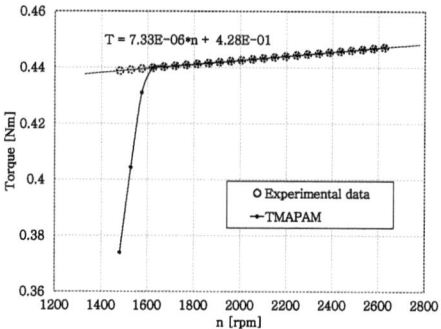

Figure 6.3.: Comparison between rotor and drive torques versus the rotational speed

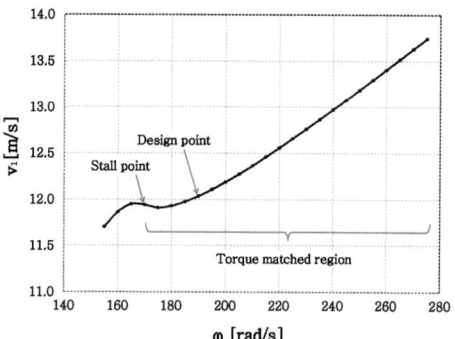

Figure 6.4.: Torque matched velocity and rotational speed relation

until it disappears at $\lambda = 3.7$, which is near the design value of 4.08 and so on for higher λ's values. In general, the distribution of α is non-uniform along the blade span except for the design λ, where the distribution of α is almost constant and equal to the design α_d.

6. Performance Analysis

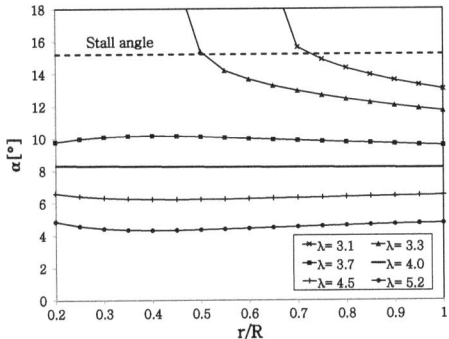

Figure 6.5.: Angle of attack (α) distribution along the rotor radius ratio for different tip speed ratios

6.2. Performance analysis of the stall-regulated WT

The analysis for prediction the performance of the stall-regulated NREL UAE phase-VI $10kW$ is mainly based on the BEM theory and it considers the post-stall region. A Matlab program is built to analyse any defined wind turbine blade shape. The information needed for the performance analysis are rotor radius R, number of blades B, number of sections, chord length distribution $c_{(r)}$, tip pitch angle β_o and twist angle distribution $\beta_{t(r)}$. In addition to the aforementioned, the operation conditions, which include wind velocity range v_1 from $5m/s$ to $25m/s$ with an interval of $1m/s$ and the associated angular speed ω, which depend on the control strategy. The angular speed is set here as a linear function of wind velocity until v_1 reaches $7.2m/s$, then it is set to be constant at $7.5rad/s$, since the control strategy of the NREL 10kW turbine is stall-regulated.

In order to calculate forces and power, BEM states that induction factors have to be evaluated as they reveal the magnitude and the direction of the air flow, which is needed for finding the relative velocity. In the BEM theory a and a' are not constant and dependent on λ_r. In the prediction procedure, the induction factors are calculated by iteration with equations (2.53) and (2.54) and continue with the procedure of BEM until the convergence. As mentioned in section Blade Element Momentum (BEM) theory, simple BEM theory breaks down as the value of a reaches greater than 0.2. Thus the correction value of a of equation (2.55) is used.

As the BEM theory does not yield reliable results for axial induction factors greater than 0.5, an additional correction by Glauert, shown in equation (6.5), to calculate a will be used in this range instead of equation (2.53) or (2.55) [67].

6. Performance Analysis

$r[m]$	$c(r)[m]$	$10 m/s$	$15 m/s$	$20 m/s$
1.510	0.711	$0.72 * 10^6$	$0.89 * 10^6$	$1.09 * 10^6$
2.343	0.627	$0.85 * 10^6$	$0.97 * 10^6$	$1.12 * 10^6$
3.185	0.542	$0.94 * 10^6$	$1.02 * 10^6$	$1.13 * 10^6$
4.023	0.457	$0.97 * 10^6$	$1.03 * 10^6$	$1.11 * 10^6$
4.780	0.381	$0.95 * 10^6$	$0.99 * 10^6$	$1.05 * 10^6$

Table 6.1.: Reynolds number distribution along the blade at $72 rpm$ rotational speed and some wind velocities for UAE phase-VI turbine [63]

$$a = \frac{18F - 20 - 3\sqrt{C_F(50 - 36F) + 12F(3F - 4)}}{36F - 50} \quad (6.5)$$

This equation is valid for $a > 0.4$, which corresponds to $C_F > 0.96$. C_F, the axial thrust coefficient, is calculated as:

$$C_F = \sigma(1-a)^2(C_L cos\varphi + C_D sin\varphi)/sin^2\varphi \quad (6.6)$$

In order to predict torque and power with BEM, C_L and C_D need to be calculated. They rely mainly on the angle of attack which is in turn a function of the wind velocity, angular speed and the control strategy of the turbine. In addition, Re and the centrifugal pumping (or the 3-D radial flow) are important parameters that effect C_L and C_D. Re distribution for a rotational speed of $72 rpm$ of the UAE phase-VI turbine is listed in Table 6.1. For simplifying the calculation, Re is selected to be $Re = 1.00 * 10^6$, which is the average value for all wind velocities from $5 m/s$ to $25 m/s$ along the whole blade, [63].

The prediction is done mainly by considering three different ranges of α. Since the control strategy of the turbine is a stall-regulated one, it is expected that the stall will appear and then dominate the blade sections as the wind velocity reaches high values because the ω will remain constant at this range. In this study, v_1 is used rather than α to define the ranges, since it has only one value for each run, whereas α changes along the blade section. The first part of predicting C_L and C_D is based on curve fitting equations of Delft 2-D data up to stall angle of 9.32 degrees [71]. This data should predict results, which are in excellent agreement with measurements over the first part of the power curve from $5 m/s$ to $8 m/s$ for a tip pitch angle of 2,3 or 4 degrees. In the prediction procedure at the tip pitch angle of 3 degrees, 2-D Delft data provides better results by implementing the equations up to a wind velocity of $9 m/s$.

The second part, after reaching the stall angle, a reasonable prediction agreement is achieved by implementing post stall UAE-derived airfoil data and Viterna equations over the wind velocity range of $9 m/s$ to $25 m/s$. UAE-derived airfoil data of Cl and

6. Performance Analysis

Cd were averaged, which are used up to the flat plate angle of 20 degrees, where the Viterna equations were inserted, which are guided by flat plate theory. Experiments have proven that the flat plate equations agree with UAE derived data averaged along the blade at angles of attack of 20 degrees or higher. The poor prediction of the Viterna model in the velocity range between $9 - 20m/s$ is because the averaged experimental data cannot help in prediction of post-stall region due to the 3-D complexities; such as producing a standing vortex in the on-board region as a result of interaction between the mean flow and the radial pumping one. Although, attention is paid when implementing Viterna equations after leading edge separation, it is taken into account that the equations are influenced by the aspect ratio of the blade whose selection dictates the maximum drag coefficient and governs the predicted power at high wind speeds over $15m/s$, it still gives poor prediction. Hence, the Viterna model is used for the velocity range of $20 - 25m/s$, and better results are achieved by adjusting the constant of C_{Dmax} while keeping the aspect ratio (AR) of 14.

$$C_{Dmax} = 1.11 + constant * AR \tag{6.7}$$

Airfoil stall angle and blade averaged values of C_L and C_D at $\alpha = 20$ degrees are used as inputs for the Viterna equations, because these values $C_{Lstall,average} = 1.24$ and $C_{Dstall} = 0.44$ conform a glide ratio for the initial conditions, which agree with the flat plate theory. Viterna equations are [71]:

$$C_L = \frac{C_{Dmax}}{2}sin(2\alpha) + (C_{Lstall} - C_{Dmax}sin(\alpha_{stall})cos(\alpha_{stall}))\frac{sin(\alpha_{stall})}{cos^2(\alpha_{stall})}\frac{cos^2(\alpha)}{sin(\alpha)} \tag{6.8}$$

$$C_D = C_{Dmax}sin^2(\alpha) + \frac{C_{Dstall} - C_{Dmax}sin^2(\alpha_{stall})}{cos(\alpha_{stall})}cos(\alpha) \tag{6.9}$$

To bridge the gap from 2-D airfoil data to the Viterna method in the wind velocity range of $10 - 19m/s$ or rather in the angle of attack range from $9.21°$ to nearly $20°$ EOLO model is implemented in this post-stall region. This model is based on flat plate equations and only valid in the transition zone between the high lift, stall development regime, dynamic stall" and the flat plate, fully stalled regime [61]. This correlates to an angle of attack range of roughly $10°$ to $20°$ according to Figure 6.6. For the fully stalled regime ($20 \leq \alpha \leq 45$), the flat plate equations are:

$$C_L = 2C_{Lmax}sin(\alpha)cos(\alpha)$$
$$C_D = C_{Dmax}sin^2(\alpha) \tag{6.10}$$

6. Performance Analysis

C_{Lmax} and C_{Dmax} are inputs for the flat plate theory equations and equal to $C_{L,\alpha} = 45°$ and $C_{D,\alpha} = 90°$ [61]. Figure 6.7 shows the trend of the experimental drag coefficient for the S809 airfoil. The value of 2.3 is assumed for this airfoil as maximum drag coefficient. Figure 6.6 shows the trend of the experimental lift coefficient for the S809 airfoil. In order to follow flat plate theory, value of C_{Lmax} at 45° would be 1.24.

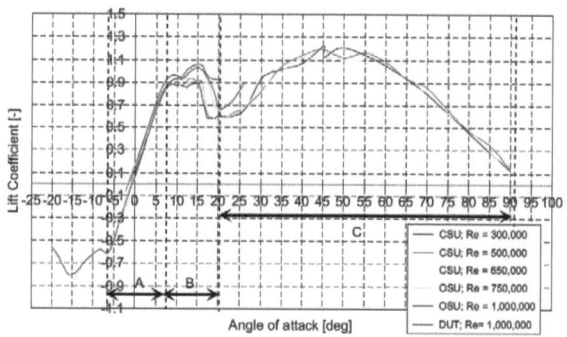

Figure 6.6.: Experimental lift coefficient of the S809 airfoil. A: "attached flow regime"; B: "high lift, stall development regime, dynamic stall"; C: "flat plate, fully stalled regime" [61].

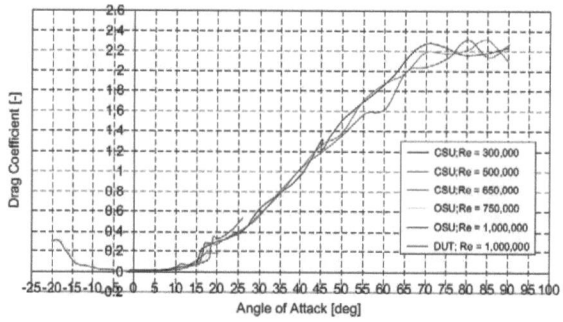

Figure 6.7.: Experimental drag coefficient of the S809 airfoil [61].

Instead of assuming one value for C_{Lmax}, EOLO model uses an increased lift coefficient to take into account the radial flow along the blades that provides an increase in energy with the same flow [61]. After some analysis, reasonable agreement was found by a nonlinear function in equation (6.11), which considers experimental results.

$$C_{Lmax} = -1.92 * 10^{-3}v_1^3 + 0.108v_1^2 - 2.05v_1 + 14.52 \tag{6.11}$$

After estimating C_L and C_D, then evaluating a and a' for all the range of v_1, all required components for calculating the force, torque and power of the turbine with equations (2.57) and (2.58) are available. Tip losses are already considered by Prandtl's tip loss correction factor and to account for profile losses along the blade equation (2.65), which include an appearing drag at higher wind velocities and angles of attack, it is multiplied by the power [47]. The performance analysis procedure diagram of the stall-regulated wind turbine is illustrated in Figure 6.8.

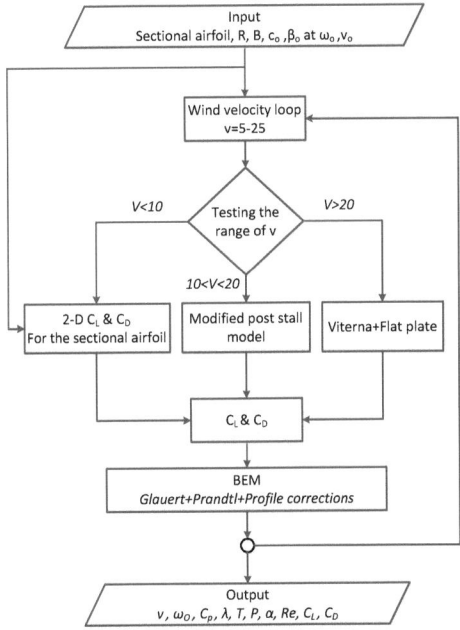

Figure 6.8.: Performance validation procedure diagram of the stall-regulated turbine.

Figure 6.9 shows the torque prediction in comparison with six further prediction models and the experimental data. Best agreement with experimental data is perceptible by the present performance analysis method, which is named "$calc$" in the Figure. From $5m/s$ to $9m/s$ all prediction models differ only slightly. A discrepancy in torque prediction of nearly $1kNm$ at wind velocity of $10m/s$ to $19m/s$ is noticeable The model EOLO provides in this region among the six prediction models the smallest differences

6. Performance Analysis

to experimental data. Therefore it is used in the analysis. From $20m/s$ to $25m/s$ only the slope of the model of Viterna & Corrigan matches with experimental data. Hence, in the analysis, the trend of the Viterna model was shifted down by modifying the *constant* to be in the range of the experimental data. In Figures 6.10 and 6.11, the axial induction factor and angle of attack over wind velocity are shown, respectively. They are calculated by the post-stall model and aero-elastic code PHATAS and compared at three different spanwise locations. Aerodynamic modelling of PHATAS is based on the BEM theory as well [80]. Differences between the models in Figure 6.10 are minimal. Shapes of the prediction models are mostly the same. Maximum deviation appears at $10m/s$, which is a about 0.04 in prediction of the axial induction factor. This is due to the post-stall model used in the present study which starts at 10 m/s. In Figure 6.11, α slopes of the prediction models differ little at a spanwise location of $r/R = 0.3$. At spanwise locations of $r/R = 0.6$ and $r/R = 0.8$ PHATAS model predicts angles of attack, which are up to 5° higher. This could be justified because validation of the axial induction factor is at each spanwise location only a little higher than the prediction of the model of PHATAS. Since α is dependent on φ, and φ is dependent on the axial and tangential induction factor and on $\lambda_{(r)}$, a higher value of the axial induction factor leads to a lower value of φ. This reduction is increased at greater spanwise locations by the factor r. As a result, α decreases with the increase of both wind velocity and span.

In Figure 6.12, angle of attack is plotted over r/R at different wind velocities. In addition to the validation, the performance-prediction method of LSWT and measuring method of local flow angles (LFA) are also shown. The LFA is measured with five-hole probes on stalks, whose length is four fifths of the chord length and they are attached to the leading edge of the blade [84]. LFA results from the vector addition of v_1 and the local tip speed u. At low wind velocities, prediction of α hardly differs between the calculated and LSWT. One reason for the increased difference of nearly 10° at $25m/s$ between both models is that LSWT considers the induced effects of the blade configuration and those from the span-wise distribution of trailing vorticity, which is distinctive at higher wind velocities [71]. The much larger values of α of the LFA method can be explained due to a direct use of the local flow angles without converting to the angle of attack because of uncertainty over how the flow angle changes between the position of the five-hole probe and the leading edge of the airfoil [84].

In the following figures, all important results of the NREL UAE phase-VI turbine are revealed. Figure 6.13 depicts the comparison between the calculated and the experimental power coefficient over wind velocity and in Figure 6.14 power coefficient is plotted over tip speed ratio. Figures show that the adopted method, which includes the modified post-stall model, is identical with the experimental data.

6. Performance Analysis

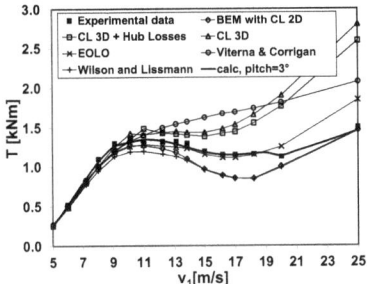

Figure 6.9.: Calculated torque in comparison with other prediction methods of the NREL UAE phase-VI rotor[61].

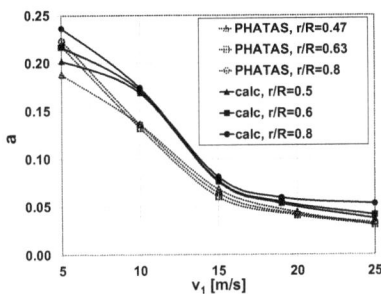

Figure 6.10.: Calculated axial induction factor and experimental data of the aero-elastic code PHATAS of the NREL UAE phase-VI rotor [80].

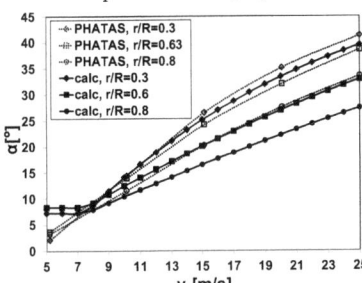

Figure 6.11.: Calculated angle of attack and experimental data of the aero-elastic code PHATAS of the NREL UAE phase-VI rotor [80].

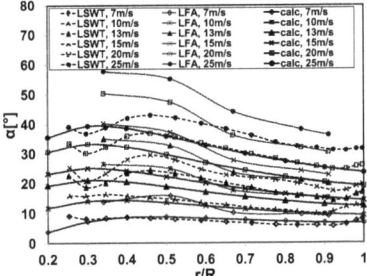

Figure 6.12.: Calculated angle of attack and prediction method of LSWT and measuring method of LFA of the NREL UAE phase-VI rotor [80].

φ and α spanwise (r/R) distribution for all wind velocities are shown in Figure 6.15 and 6.16 respectively. φ increases continuously with increasing of wind velocity. This is because φ is dependent on $1/\lambda_{(r)}$ and hence v_1, equation (2.45), thus, increasing v_1 with constant angular speed results in an increment of the relative velocity and thus φ and α.

In Figure 6.17 the axial induction factor is plotted over blade length for all wind velocities. Greatest values of a are achieved at the tip and a is continuously decreasing with wind velocity. It reveals that the higher rotor impediment appears at lower wind velocities in a stall regulated turbines, since the rotational speed is constant, which is consistent with Figure 6.10. At $7m/s$, the value of a is at its maximum and closer to the

6. Performance Analysis

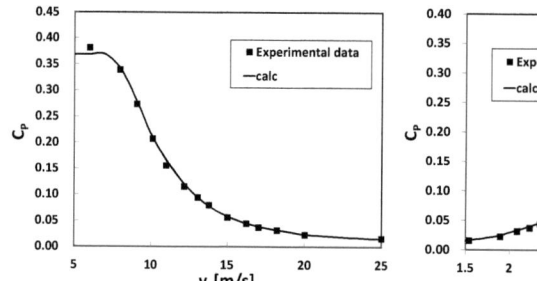

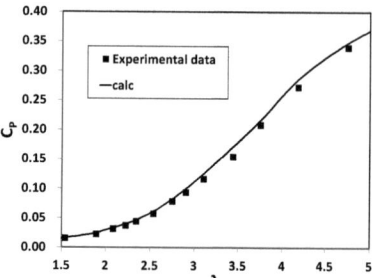

Figure 6.13.: Comparison between the calculated and the experimental power coefficient over wind velocity of the NREL UAE phase-VI rotor [61].

Figure 6.14.: Comparison between the calculated and the experimental power coefficient over tip speed ratio of the NREL UAE phase-VI rotor [61].

ideal of 1/3. Thus, the maximum power coefficient of 0.36 can be obtained, as shown in Figure 6.13.

The tangential induction factor is shown in Figure 6.18. Values of a' decrease from the hub to the tip to nearly zero. The maximum values near the hub are due to the higher chord length and the lower tangential speed. In general, the values of a' increase with the v_1 which indicates more tangential deviation or more wind-rotor intersection, since ω is constant for the stall-regulated wind turbines.

6. Performance Analysis

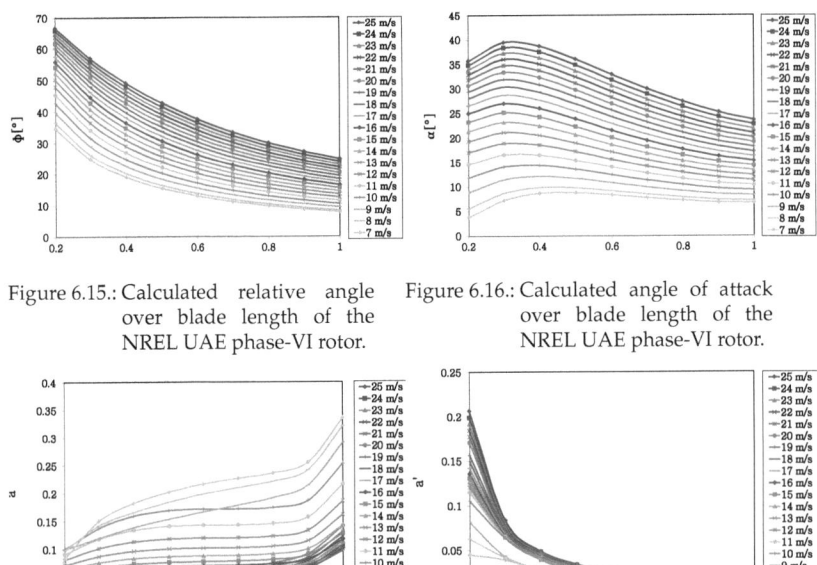

Figure 6.15.: Calculated relative angle over blade length of the NREL UAE phase-VI rotor.

Figure 6.16.: Calculated angle of attack over blade length of the NREL UAE phase-VI rotor.

Figure 6.17.: Calculated axial induction factor over blade length and wind velocity of the NREL UAE phase-VI rotor.

Figure 6.18.: Calculated tangential induction factor over blade length and wind velocity of the NREL UAE phase-VI rotor.

6.3. Modified Post-stall model

As indicated above, the Viterna model with some modifications is performing better in the higher wind velocity range $v_1 \geq 20$, whereas the EOLO model can predict the performance within the stall region $9 < v_1 < 20$. They both serve to predict the performance of the baseline turbine of NREL phase-VI 10kW. In order to make EOLO perform for different scales and shapes of turbines, it has to be modified. A semi-empirical equation, which is derived from EOLO and the existing experimental data with additional observations is developed in the present study to overcome the performance prediction of any existing wind turbine. All parameters defining the turbine rotor, such as rotor radius, chord, the number of blades are included in the developed model.

$$k_1 = \left[-0.0003 \left(\frac{R}{R_i}\right)^2 + 0.0126 \left(\frac{R}{R_i}\right) + 0.375\right] \left(\frac{R}{R_i}\right)^{0.1} \left(\frac{c_i}{c}\right)^{\frac{c_i}{4c}} \quad (6.12)$$

$$k_2 = \left[0.0006 \left(\frac{R}{R_i}\right) + 0.011\right] \left(\frac{R}{R_i}\right)^{-0.25} \left(\frac{c_i}{c}\right)^{\frac{c_i}{4c}} \quad (6.13)$$

$$k_3 = \left[k_2 \left(\frac{R}{R_i}\right)\right]^{\frac{R_i}{R}} \quad (6.14)$$

$$k_4 = -0.57 \, log(k_3) + k_1 \quad (6.15)$$

$$k_5 = 2.4 \, C_{L,max} - 2 \quad (6.16)$$

where, the suffix i refers to reference wind turbine parameters, which are the radius $R_i = 5 \, m$ and the mean chord $c_i = 0.5606$. These constants used to calculate the maximum lift coefficient as

$$C_{L,max} = \left[\frac{k_3}{sin(v_1)}\right]^{k_4} + k5 \quad (6.17)$$

then use it in flat plate equations 6.10 to get C_L and C_D.

Figures 6.19, 6.20, 6.21 below reveal the advantage of using the model over the equation (the model refers to the modified model and the equation is the equation used to estimate EOLO, equation 6.10) for the performance prediction of different turbine characteristics. The main criteria used to measure the advantage are limit values of the post-stall region and trends of the curves. Limit values are simply characterized by the starting value of the stall at $v_1 < 10$. This is trusted since it is found by 2D Delft equations and the end value at $v_1 = 20$, which is calculated by the Viterna model, which is trusted for wind velocity at that range since the airfoil will behave like a flat plate as explained above. The second criterion is the trend of the curve, where the standard trend is taken from the available experimental data, which shows a continued decrement of the torque after $v_1 = 10$ until reaching the end value of $v_1 = 20$. In addition, the smoothness of the curve can be another indication for the feasibility of the model.

It is required to mention here that B is not set in the model since it can be predicted rather well by both the model and the equation. Nevertheless, the model still shows better trends especially for the torque in Figure 6.19. Figure 6.20 shows the torque versus the wind velocity of the model in comparison with the equation for a different chord ratios, which is the ratio of any wind turbine mean chord c by the baseline turbine mean chord (c_i). Here, $c_i = 0.5606$. It is found that the model is applicable until $c/c_i = 1.75$, and it can predict power and torque in much better way than the equation. Hence it is included in the model, equation 6.16.

6. Performance Analysis

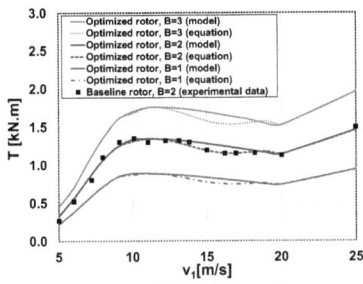

Figure 6.19.: Model and equation torque for different number of blades (B=1,2,3).

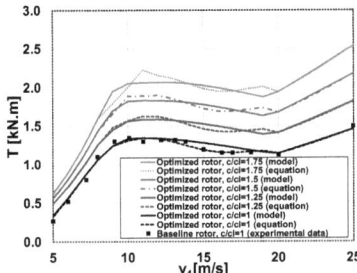

Figure 6.20.: Model and equation torque for different chord ratios ($c/c_i = 1, 1.25, 1.5, 1.75$).

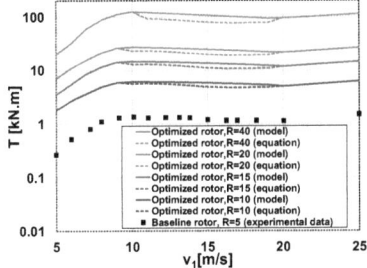

Figure 6.21.: Model and equation torque for different turbine radius ($R = 5, 10, 15, 20, 40m$).

Figure 6.21 shows the effect of the rotor radius on the torque for both model and equation. The model is applicable until radius value of 40 m, and for all of the radius ranges, it performs better than the equation. Consequently, the torque increases with the increment of the rotor radius, and hence the wind power increases too (for the new turbine area), and the angular velocity is kept constant, thus the power coefficient decreases. Since, the model is set based on the information obtained from the baseline turbine; for instance, in equation 6.17 it is set in terms of wind velocity rather than angle of attack, the relation between α and v_1 has to be checked for better prediction at different radius values. As the radius increases, the tangential velocity increases, and hence the angle of attack decreases. Figure 6.22 proves that fact, where it reveals that within the tested radius range, there is a small change (reduction) of the mean angle of attack as the wind velocity increases, hence, the stall will delay as the rotor radius increases. Thus, the range of wind velocity defining the post-stall range has to be shifted as the radius increases. It is depicted in Figure 6.22 with dotted lines, for the

6. Performance Analysis

same mean angle of attack where the stall starts about $\alpha_{mean} = 10$, which is associated with wind velocities $v_1 = 9.5, 10, 10.5, 11$ of the radius $R = 5, 10, 20, 40$, respectively. The stall delay with the radius increase is depicted in Figure 6.23, where the locus of the different radii torques form an exponential curve.

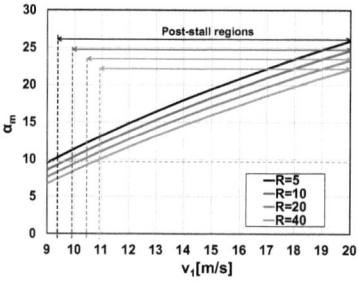

Figure 6.22.: Mean angle of attack versus wind velocity for different radius.

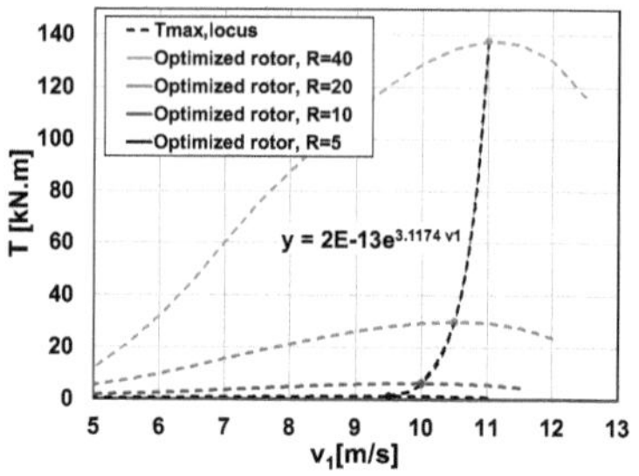

Figure 6.23.: Effect of radius on the torque over wind velocities.

6.4. Performance prediction of the stall-regulated shape optimized WT

In the performance prediction of the shape optimized turbine, the original chord length and the twist angle distributions of the baseline turbine are replaced by equations from Figure 5.10 of the optimized blade shape. The BEM method with the modified stall model are used in the analysis to resolve the increment of the chord length in the shape optimized turbine.

In Figure 6.24, the power prediction of the shape optimized turbine is higher than the experimental data of the baseline turbine and the velocity decreases from $v_{1,o} = 7.2m/s$ to the shape optimized turbine velocity of $v_1 = 6.85m/s$. In the post-stall region $9 > v_1 > 20m/s$, the rated power is increased from $P_{N,o} = 10$ to $P_N = 14.5kW$. This is due to the increase in the torque from $T_{N,o} = 1.33[N.m]$ to $T_N = 2[N.m]$, while the angular speed is kept constant at $\omega_o = 7.5 rad/s$. Hence, the post-stall model that includes the increase of the chord is performing well as the mean chord increment is limited to a value less than 2 of the baseline turbine mean chord. In Figure 6.25, the calculated power coefficient is plotted over tip speed ratio. It is recognizable in comparison with the experimental data of the baseline turbine that C_P is raised for the optimized blade shape. At the baseline turbine design tip speed ratio of $\lambda_o = 5.2$ the C_P is increased

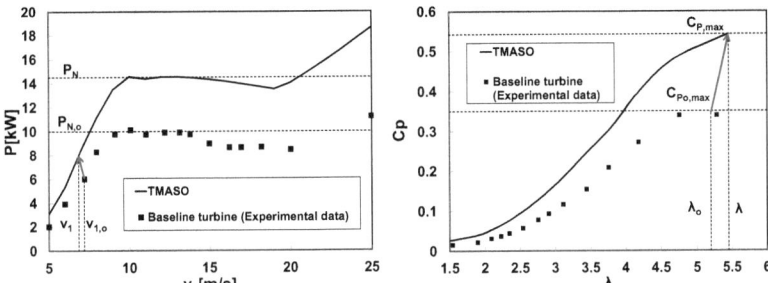

Figure 6.24.: Power increment of the shape optimized turbine as compared to the baseline turbine experimental data [61].

Figure 6.25.: Power coefficient increment of the shape optimized turbine as compared to the baseline turbine experimental data [61].

from about 0.35 to 0.54 at the new optimized tip speed ratio of $\lambda = 5.45$. As a result, the objective to increase the power coefficient by an optimized blade shape is achieved.

6.5. Performance Analysis of the Pitch-Control Optimized HAWT

This section includes the performance analysis results of pitch-control optimized NREL UAE phase-VI wind turbine. The pitch-control angle distribution has a great affect on the prediction procedure. Little changes cause large amplitudes. By the pitch optimization procedure the best possible distribution was found. In comparison to the stall-regulated procedure in section 6.2, the pitch control angle $\beta_{c(v_1)}$ distribution in equation (5.21) is added to the twist distribution in equation (5.18) in the range of wind velocities greater than the rated one. Additionally, curve fitting equations of the lift and drag coefficients of Delft 2-D data are used along the whole range of wind velocities. That is the case because angles of attack greater than the stall angle of $9.32°$ should not be reached. This is achieved by adding the pitch control angle $\beta_{c(v_1)}$ distribution to the collective pitch distribution $\beta = \beta_{t(r)} + \beta_o$. This, in turn, leads to a decrease in angle of attack ($\alpha = \varphi - (\beta + \beta_c)$). Results show that α increases with the increase of wind velocity and it reaches a range of values of about $9.5 - 14.5°$ at v_1 of $10 m/s$, illustrated in Figure 6.26. Consequently, at $12 m/s$, the angle of attack is reduced to a value of maximum $9°$. From this point on, the maximum value of α increases until $20 m/s$ to roughly $13°$. Compared to the distribution of α without pitch control optimization in Figure 6.16, the angle of attack is reduced considerably.

6. *Performance Analysis*

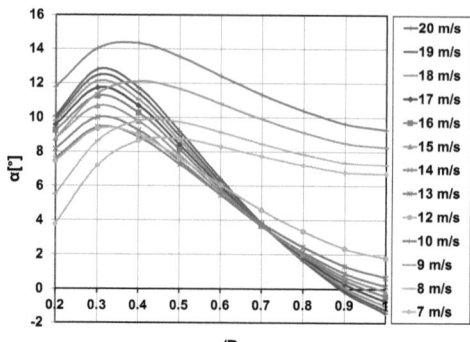

Figure 6.26.: Calculated angle of attack over blade length of the pitch optimized NREL UAE phase-VI rotor.

Calculation and results of $\omega - v_1$ and $\varphi - r/R$ do not vary from section Performance analysis of the stall-regulated WT and are, therefore, not further described. Calculation of axial and tangential induction factors is done with the same procedure as in the stall-regulated one, section 6.2. Only the calculation of lift and drag coefficients differs. In section 6.2, the prediction of C_L and C_D is based on three different models in three different ranges of wind velocity. Because of the pitch control strategy the angle of attack range will always be less than the stall angle thus the post-stall model is not included here. Hence, the C_L and C_D curve fitting equations of the sectional airfoil are used for the whole range of wind velocities. The procedure diagram of the pitch-control validation is shown in Figure 6.27. In comparison to the validation of the stall-regulated turbine, calculation of C_P over v_1 and λ are hardly different.

Shape of power prediction is shown in Figure 6.28. It reveals the increment of the power at the wind velocity range higher than the rated one as compared to the stall-regulated control. The pitch control angle distribution perfectly performs at the range of wind velocity $v_1 \geq 10 m/s$. In Figure 6.29, C_P is higher than in Figure 6.14 and lies in the range of tip speed ratio between about 2 and 4, which corresponds to wind velocities between $10 m/s$ and $20 m/s$ in Figure 6.28. This is because the pitch-control strategy can extract more energy from the oncoming wind velocity at the range greater than the rated velocity, which lies at about $10 m/s$, as explained before in the Wind Turbines Control Strategies section.

6. Performance Analysis

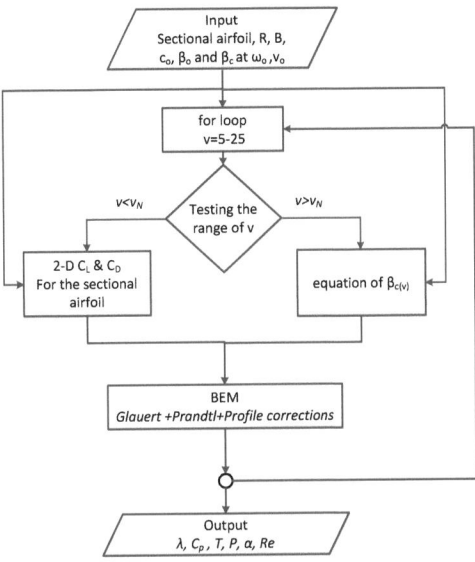

Figure 6.27.: Pitch-control performance analysis diagram.

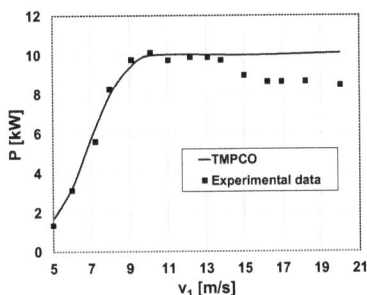

Figure 6.28.: Comparison between the calculated and the experimental power of the pitch optimized NREL UAE phase-VI rotor [61].

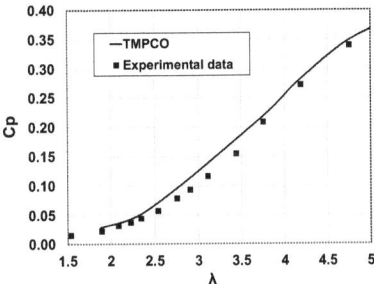

Figure 6.29.: Comparison between the calculated and the experimental power coefficient over tip speed ratio of the pitch optimized NREL UAE phase-VI rotor [61].

6. Performance Analysis

6.6. NREL 5MW Wind Turbine

To verify the functionality of the analysis procedure of section 6.5 for different turbines scales, properties of the NREL 5MW wind turbine, called $REpower5M$ [55], are used as inputs. The NREL 5MW wind turbine was chosen because of the detailed information available. It is also representative of typical utility-scale land- and sea-based multimegawatt turbines [55]. Specifications of the NREL 5MW turbine are described in appendix B.1.

6.6.1. Performance analysis of NREL 5MW

The performance analysis is mainly based on BEM theory as it is done for the NREL 10kW. Most of the differences to be considered are the changed turbine properties and control strategy.

The radius of the three bladed rotor is shortened from $R = 61.6333m$ to $R = 61m$ to facilitate the calculations. The number of sections of the blade is set to 61. The wind velocity range v_1 from $5m/s$ to $25m/s$, where the last is the cut-out wind velocity, with an interval of $1m/s$ is kept the same. Rated rotational speed gives a rated angular speed of $\omega = 1.267 rad/s$. After reaching $v_1 = 11.4m/s$ rated angular speed is constant at $1.267 rad/s$. The chord length and twist angle distributions are replaced by curve fitting equations for Figures B.2 and B.3.

Due to the pitch control strategy, the pitch control angle distribution $\beta_c(v_1)$ has to be added to the twist distribution $\beta_t(r)$. Shape of $\beta_c(v_1)$ is depicted in Figure B.5 and in equation 6.18. The pitch angle increases after reaching the rated wind velocity of $v_1 = 11.4m/s$ to reduce the angle of attack as the wind velocity increases more than the rated one to prevent the stall and hence to keep the rated power constant.

$$\begin{aligned}\beta_{c(v_1)}[°] = &- 7.005 * 10^{-5}v_1^6 + 7.94 * 10^{-3}v_1^5 - 0.372v_1^4 \\ &+ 9.18v_1^3 - 126.4v_1^2 + 920.9v_1 - 2775.9\end{aligned} \quad (6.18)$$

Calculations of axial and tangential induction factors are done with the same procedure as in the stall-regulated one. Whereas, the calculation of lift and drag coefficients are done as explained in the pitch-control one, section Performance Analysis of the Pitch-Control Optimized HAWT. The C_L and C_D curve fitting equations of the six airfoils profiles are used for the whole range of wind velocities. Profiles of $Cylinder1$ and $Cylinder2$ are neglected because their lift coefficient is zero [55]. Hence, prediction of C_L and C_D starts with profile $DU40\_A17$ in this procedure. Referring to the location of the airfoil profiles in table B.1, a practical grate for dividing the blade is defined in Figure 6.30.

6. Performance Analysis

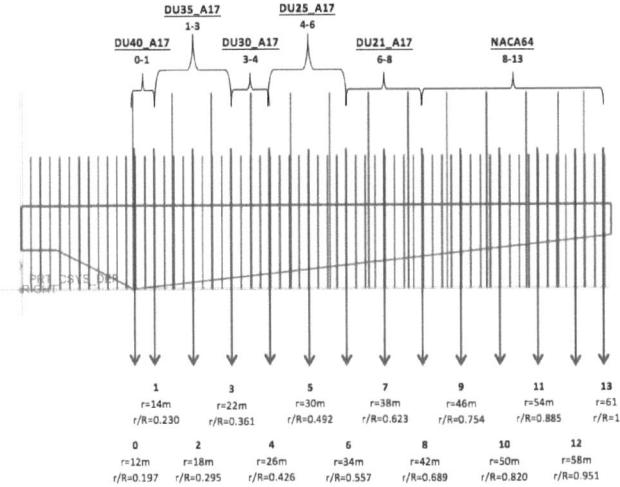

Figure 6.30.: Airfoil profiles and sections of the NREL $5MW$ wind turbine from $r = 12m$ to the tip.

Defining the sections to 61, it is barely possible to separate the start and end location of a profile by one meter. Thus, profile $DU40\_A17$ begins at $r = 12m$ instead of $r = 11.7500m$, relate to table B.1. Each profile has its spanwise location center at the whole-number rounded $RNodes$ and its start and end point at half of the $DRNodes$ value less and more than the value of the $Rnodes$. The $DRNodes$ value of $4.1000m$ in table B.1 is reduced to $4m$. Calculation of power and torque are conducted by the same method as in section Performance Analysis of the Pitch-Control Optimized HAWT. In Figure 6.31 experimental data and calculated power is plotted over wind velocity. Shapes of the power prediction is in a perfect agreement with the experimental data, which is obtained from [55].

In Figure 6.32, the angle of attack is plotted over blade length for different wind velocities. After reaching rated wind velocity, it is perceptible that α is reduced by the pitch control angle distribution ($\alpha = \varphi - (\beta + \beta_c)$). Values of the distribution of φ in Figure 6.33 are a little smaller than the values of the $10kW$ distribution of φ in Figure 6.15 because of a larger radius of the $5MW$ turbine according to equation (2.45), for the same approximately induction factors, as will be shown in later figures. Drop in α results in a decline in C_L and C_D, Figures 6.34 and 6.35, and drop in the values of C_L and C_D with the increasing of ϕ result in a decrease of a, Figure 6.36, equation (2.53). a'

6. Performance Analysis

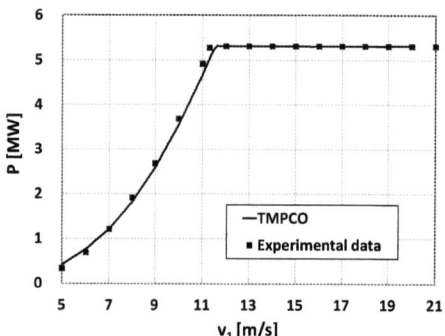

Figure 6.31.: Comparison between the calculated and the experimental power of the NREL $5MW$ wind turbine [55].

remain nearly constant over the wind velocity range in Figure 6.37.

Figure 6.38 depicts the trend of the power coefficient over wind velocity and in Figure 6.39, the power coefficient is plotted over tip speed ratio. In both figures, experimental data is used from the DOWEC project, which considered a $6MW$ turbine with a rotor diameter of $129m$, a rated wind velocity of $12.1m/s$ and a rated rotational speed of $11.844rpm$ [60]. Any other properties as used profiles, chord length and twist angle distribution and pitch regulation are the same [60]. Hence, both shapes agree well with experimental data under the assumption that the trend of the prediction of C_P over v_1 and C_P over λ should be slightly smaller.

6. Performance Analysis

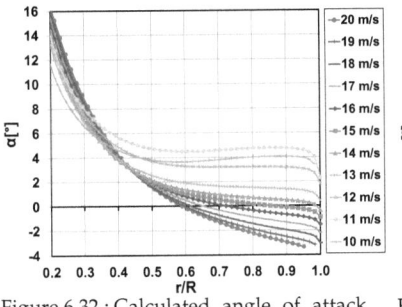

Figure 6.32.: Calculated angle of attack over blade length of the NREL $5MW$ wind turbine.

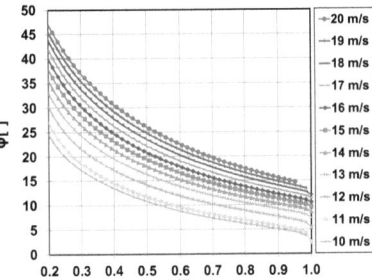

Figure 6.33.: Calculated phi over blade length of the NREL $5MW$ wind turbine.

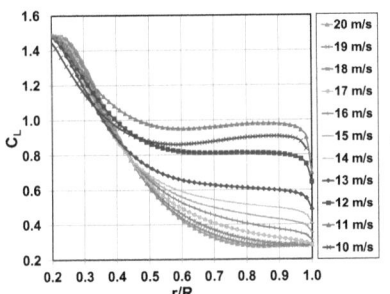

Figure 6.34.: Calculated lift coefficient over blade length and wind velocity of the NREL $5MW$ wind turbine.

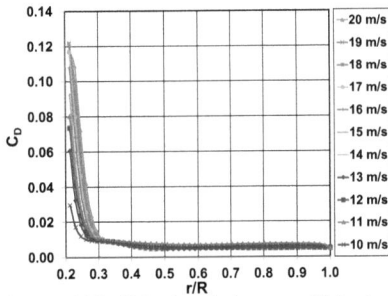

Figure 6.35.: Calculated drag coefficient over blade length and wind velocity of the NREL $5MW$ wind turbine.

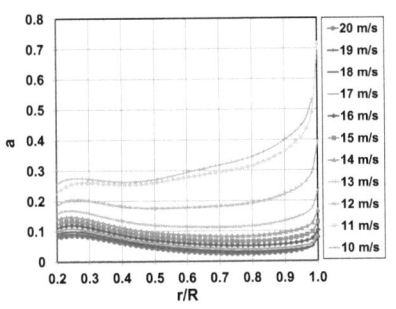

Figure 6.36.: Calculated axial induction factor over blade length and wind velocity of the NREL $5MW$ wind turbine.

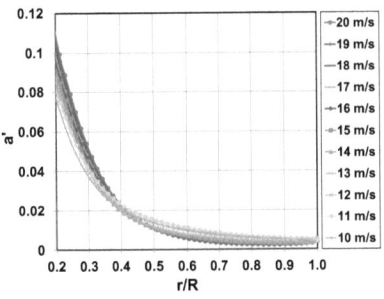

Figure 6.37.: Calculated tangential induction factor over blade length and wind velocity of the NREL $5MW$ wind turbine.

6. *Performance Analysis*

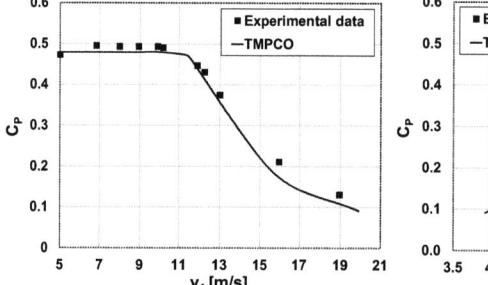

Figure 6.38.: Comparison between the calculated and the experimental power coefficient over wind velocity of the NREL 6MW wind turbine [60].

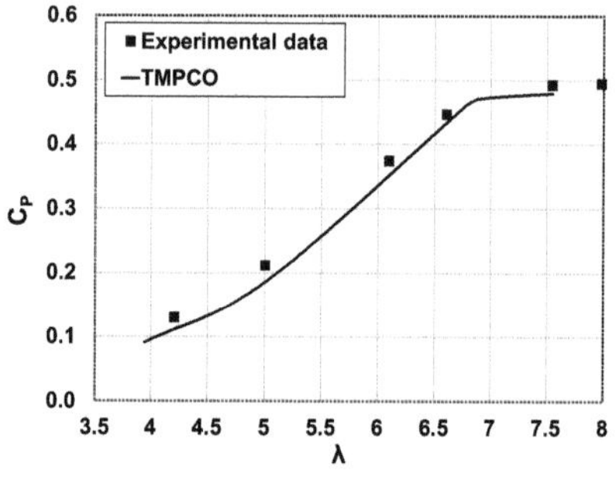

Figure 6.39.: Comparison between the calculated and the experimental power coefficient over tip speed ratio of the NREL 6MW wind turbine[55, 60].

7

Numerical Validations

In this chapter, the wind turbine that optimized with using of TMASO method (chapter 5) and analysed with TMAPAM method (chapter 6) is validated by numerical simulation analysis. Further numerical investigations are conducted to predict the flow separation at high angles of attack and the surface streamlines. Additional investigations for the control strategies are presented later in this chapter.

7.1. Power Coefficient

The power coefficient is plotted over the tip speed ratio for the simulations in comparison to the provided experimental data in Figure 7.1.

The simulated points in the design region are in good agreement with the experimental data. The design region is in the tip speed ratio region of $3.9 < \lambda < 4.5$. The design point is at $\lambda = 4.08$, which corresponds to the design angle of attack $\alpha_{design} = 8°$. A decrease in λ causes an increase in the angle of attack α along the whole blade span. In the low angle of attack region the simulations still in its good agreement and reflect the same experimental trend. In the high angle of attack region one data point is provided, which indicates a sharp decrease of power coefficient, due to the beginning of stall. In the simulations a more gentle decrease in power coefficient is seen. The simulations were performed under the assumption of mainly steady flow phenomena. The

7. Numerical Validations

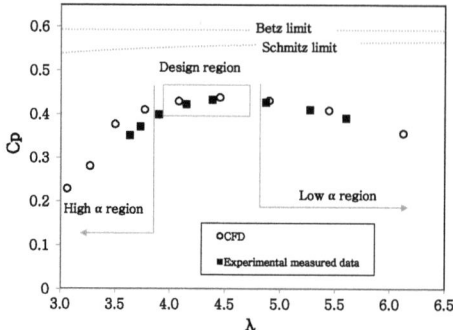

Figure 7.1.: Power coefficient over tip speed ratio

flow in separated flow regions is highly unsteady and thus the steady RANS model may not be appropriate to investigate these states.

7.1.1. Tangential Velocity on 2-D Blade Sections

Computational Fluid Dynamics simulations offer extensive insights to the computed flow. Here the flow over the blade on representative sections is examined. The sections were chosen as:

- r=0.3R: The first airfoil profile of the blade is at r=0.2R. At this section root and rotational effects are expected to play a role

- r=0.6R: The middle of the aerodynamic profile. The flow here is largely of two dimensional character

- r=0.9R: Approximately up to this section influences of the tip vortex are expected to influence the flow over the blade

The tangential component of the velocity, seen from the rotating blade, is shown along velocity streamlines in Figure 7.2 for low (a)(b), design (c), and high (e)(f) angle of attack.

For the low and design angle of attack configuration the flow over the blade seems largely undisturbed at all sections. However there is a reversed boundary layer flow at the trailing edge. As can be seen by the legends, the maximum velocity of the reversal flow increases with increasing angle of attack. For the first high angle of attack configuration at $\lambda = 3.5$ there is clear separation at the mid section, whereas the root

7. Numerical Validations

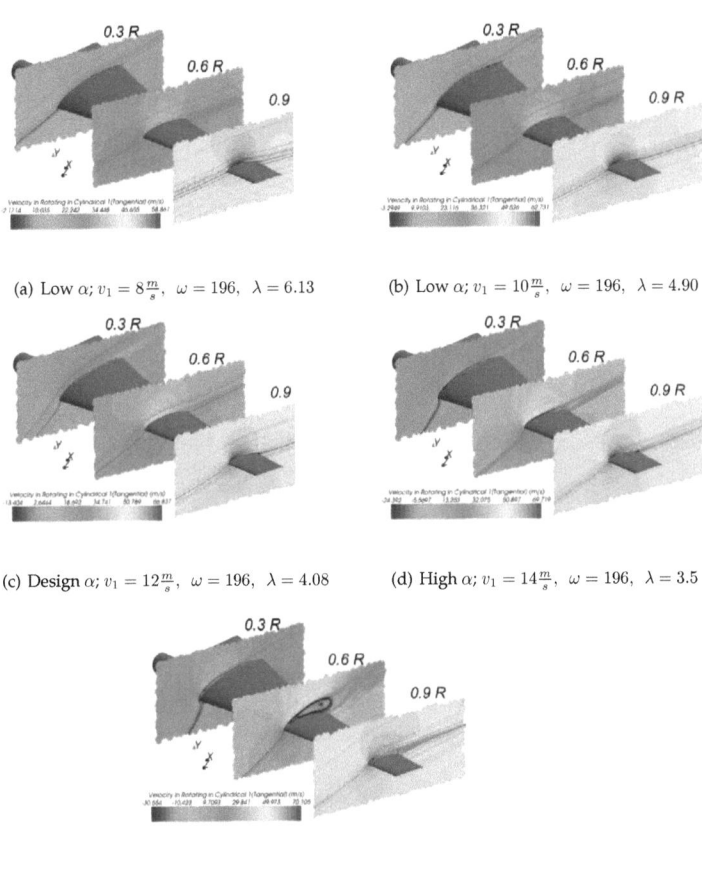

Figure 7.2.: Tangential velocity on 2-D blade sections with velocity streamlines at low, design and high angle of attack

section seem to be not affected and the tip section something in between. For the tip speed ratio of $\lambda = 3.06$ the flow on the mid span suction side is detached completely. On the tip region there also is large separation. The root section shows a separation bubble near the leading edge, but no separation at the trailing edge. Due to large separated areas the applicability of the RANS model has to be questioned for the second high angle of attack simulation. Figure 7.3 shows the residual of the turbulent kinetic

7. Numerical Validations

energy (TKE) over the tip speed ratio. With decreasing tip speed ratio, thus increasing angle of attack, the residual for the TKE increases. At $\lambda = 3.5$ the course of the TKE turns abruptly steep. Here the airfoil flow stalls and the applied numerical models are no longer appropriate. Thus the simulations with $\lambda < 3.5$ at design pitch angle and design angular velocity are no longer considered.

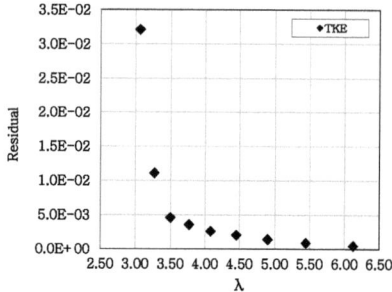

Figure 7.3.: Residual for turbulent kinetic energy over tip speed ratio λ

7.1.2. Pressure Coefficient Distribution on Sections

The flow over the three sections along the blade are at similar angle of attack and Reynolds number. In Figure 7.4 the non dimensional pressure coefficient is plotted for the sections over the non dimensional chord length. With this non dimensional treatment the flow over the airfoil sections is compared. It catches the eye, that the flow at the different sections differs for the compared angles of attack. Thus, the flow is not only dependent on the angle of attack and Reynolds number but also on the position along the blade span. Further there is no obvious pattern for the difference in pressure coefficient course between the sections, that applies to all tested angles of attack. The integral over the closed pressure coefficient course is the lift coefficient of the section. So the larger the included area, the higher the lift. The drag cannot solely be determined from the pressure coefficient, however indications are given, e.g. for separated flows, that increase drag. For the tip speed ratio of $\lambda = 6.13$ the C_p trend on the pressure side (bottom) is similar for the sections. The C_p value of 1 at $x/c = 0$ marks the stagnation point at the leading edge. On the suction side (top) the root section is constantly below the middle and tip section, which are similar. For the root section the value of C_p at $x/c = 1$ is zero. For a sharp trailing edge, as in the simulations, this indicates an attached flow at the trailing edge. At the middle and tip section the pressure coefficient there is slightly negative, indicating a separated flow

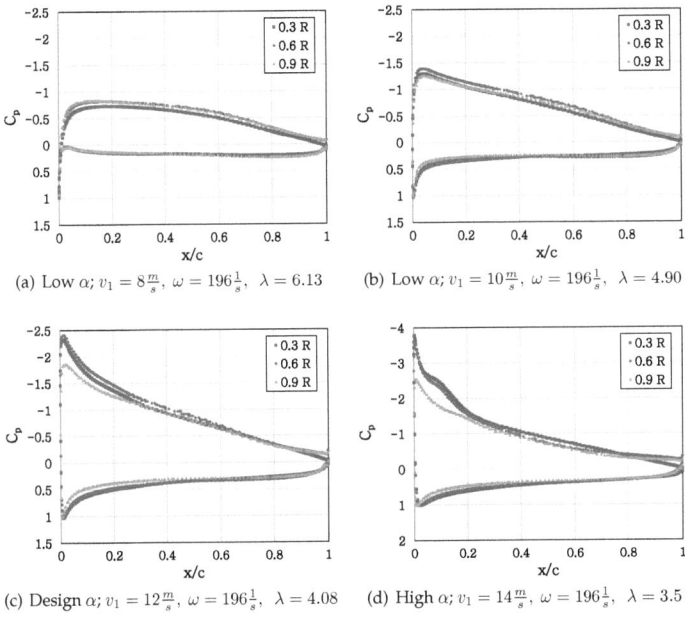

Figure 7.4.: Pressure coefficient on blade sections

at the trailing edge suction side. This trailing edge behaviour for the three sections qualitatively can be seen at all tested angles of attack. At the design configuration at $\lambda = 4.08$ the pressure coefficient course near the leading edge is similar for the root and middle section, whereas the corresponding total values at the tip section are lower. The low angle of attack configuration at $\lambda = 4.90$ can be seen as a blend between the two previously analysed states. At the high angle of attack of $\lambda = 3.50$ the C_P distribution differs from the design configuration in two main points. Near the leading edge on the suction side the pressure coefficient drops steep, followed by a flatter decrease. This drop indicates a separated bubble on the root and middle section. The tip section does not show this steep decrease. Secondly the flat course of the middle and tip section from approximately $x/c = 0.7$ to the trailing edge indicates large areas of separated flow near the trailing edge.

7.1.3. 3-D Flow on Blade

For a better understanding of the differences between the flow over the investigated sections the three dimensional flow on the blade surface is analyzed. Figure 7.5 shows streamlines alongside the pressure on the suction (S) and pressure (P) sides of the blade, for the low angle of attack $\lambda = 6.13$ (a), the design configuration $\lambda = 4.08$ (b) and the high angle of attack $\lambda = 3.5$ (c).

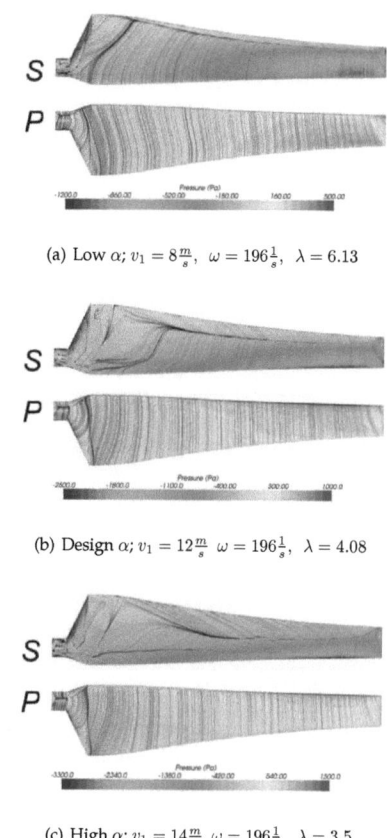

(a) Low α; $v_1 = 8\frac{m}{s}$, $\omega = 196\frac{1}{s}$, $\lambda = 6.13$

(b) Design α; $v_1 = 12\frac{m}{s}$ $\omega = 196\frac{1}{s}$, $\lambda = 4.08$

(c) High α; $v_1 = 14\frac{m}{s}$ $\omega = 196\frac{1}{s}$, $\lambda = 3.5$

Figure 7.5.: Pressure distribution and blade streamlines at different angles of attack

For all angles of attack the pressure side flow is mainly of two dimensional character, apart from the tip near region. The suction side flow at low angle of attack in subfigure (a) shows reversal boundary layer flow at the trailing edge apart from the root near section and a small part next to the tip. The tangential flow direction can be distinguished in the plots by considering the effect of centrifugal acceleration on the boundary layer flow. An axial velocity component, that points towards the root, when considering a chord-wise flow, is a clear indication of a reversed boundary layer flow and thus separation. At the design angle of attack the separated strip at the trailing edge is wider than at low α. The root near section is characterized by radial velocity components, which originate from the rotational effects, namely Coriolis and centrifugal acceleration, as discussed previously. At high angle of attack the trailing edge separation is largely increased. In the span-wise middle section of the blade the flow is separated over approximately half the chord length. On the suction side separation also occurs near the leading edge. The flow is separated and reattaches, thus forming a separation bubble [69]. For the tip near flow this separation bubble does not form. The non existing trailing edge separation at the root near section agrees with the theory of suppressed boundary layer separation near the root, through Coriolis and centrifugal forces, as outlined by Shen [83]. For the design and high angle of attack configurations the flow is not solely in chord-wise direction, but of three dimensional character, for large parts of the blade.

7.2. VS-VP analysis

In this section the turbine power related values and the flow over the blade for the variable speed - variable pitch (VS-VP) strategy is analysed. The intention is to provide information for the optimization code TMPCO that suggests a pitch control for the HAWT for a range of wind velocities. The simulations were performed with the variable pitch model.

7.2.1. VS-VP HAWT Performance

For the VS-VP strategy the velocity range of $8 - 16 m/s$ was investigated. The design wind velocity is $12 m/s$, thus the range of $13 - 16 m/s$, which is over the rated velocity, is important for revealing the rated power and to validate the control pitch angle β_c that suggested by the TMPCO. The rated power was set to $100W$. As the optimization derived β_c led to power values above the rated power of $100W$, for wind velocities from 14 to $16m/s$, the β_c there was altered. To reduce the power the control pitch angle was increased. With the variable pitch model the control pitch angles were determined in an iterative process. Figure 7.6 shows the angular velocity ω and control pitch angle

β_c in (a) and C_P and power in (b).

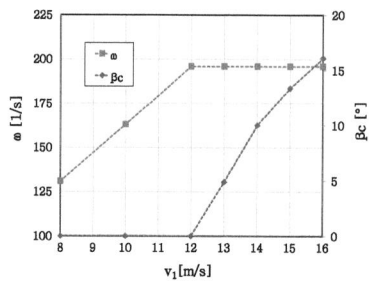
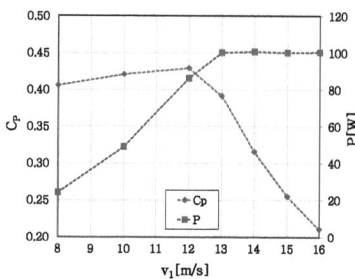

(a) Angular velocity and control pitch angle over wind velocity

(b) Power coefficient and power over wind velocity

Figure 7.6.: VS-VP control strategy

7.2.2. 3-D Flow on Blade of VS-VP

Figure 7.7 shows the streamlines on the blade alongside the pressure for the VS-VP strategy. The configuration at $v_1 = 12\frac{m}{s}$ is the same as for the FS-FP approach and is shown in Figure 7.5 (b). Up to a wind velocity of $v_1 = 12m/s$ the angle of attack is α_{design} and the main difference is the Reynolds number. The pressure sides have a two dimensional character, except for little tip influence. On the suction side the configurations show separation on the trailing edge middle and tip section. This separated area decreases with increased wind velocity and thus angular velocity and Reynolds number. An increase in Reynolds number decreases the trailing edge separation. For wind velocities from 13 to $16m/s$ the plots show a thin stripe of suction side trailing edge separation. The root near area is not affected, due to rotational effects. With increasing wind velocity the control pitch angle is increased, by so much, that the angle of attack decreases despite the increased wind velocity. Two effects can be seen; first, the locations of maximum pressure in the tip region moves from the leading edge in chord-wise direction with decreasing angle of attack. This can be seen on both, the suction and pressure, sides. Secondly, starting at $15m/s$ the pressure side tip region shows separation on the leading edge tip, that increases in span-wise direction with decreasing α.

7. Numerical Validations

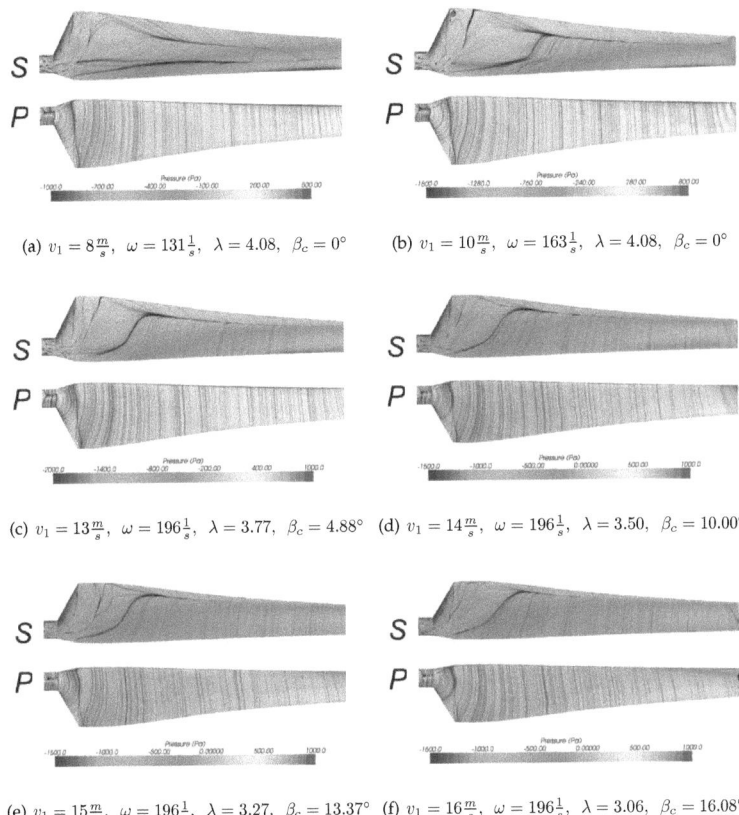

Figure 7.7.: Pressure distribution and blade streamlines for VS-VP at wind velocities $8 - 16 m/s$

7.2.3. Pressure Coefficient Distribution for VS-VP

Figure 7.8 shows the pressure coefficient over representative sections for the VS-VP approach. The separated trailing edge flow at middle and tip section can be seen for all states. The configurations at $\lambda_{design} = 4.08$ are similar to the design point, discussed in section 7.1.2. For the wind velocity range of 13 to $16 m/s$ the lift, that is related to the area enclosed by the pressure coefficient, decreases with increasing wind velocity. The control pitch angle is increased with increasing wind velocity in order to decrease the angle of attack. Differences between the span-wise sections of one simulation increase

7. Numerical Validations

with increasing control pitch angle. The root section shows the highest total values of pressure coefficient near the leading edge and the tip section the lowest. For high control pitch angle at 15 and $16 m/s$ the tip section pressure coefficient near the leading edge shows a negative lift at the tip section leading edge.

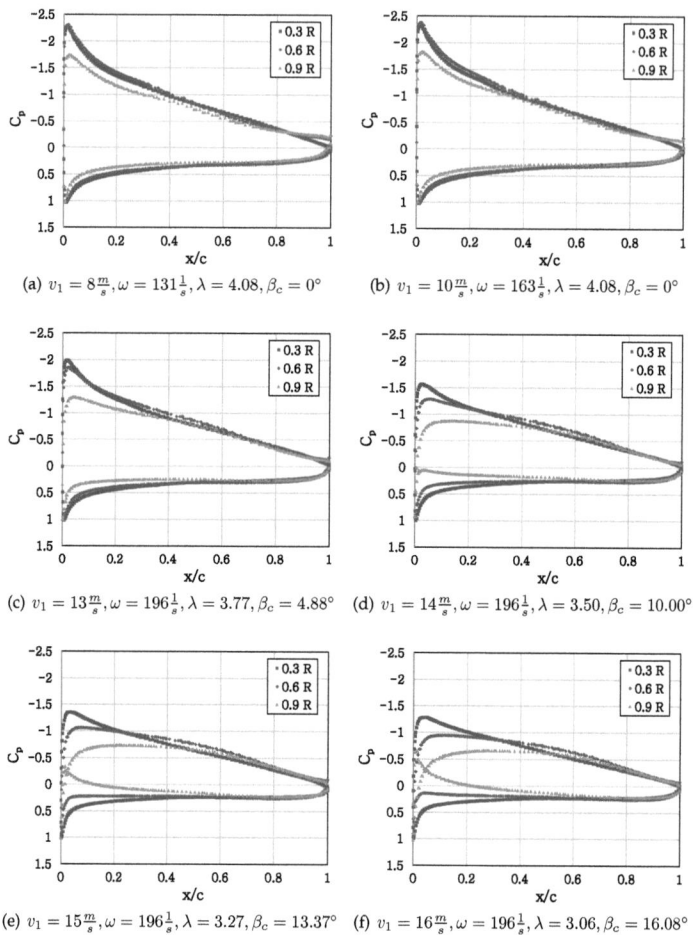

(a) $v_1 = 8\frac{m}{s}, \omega = 131\frac{1}{s}, \lambda = 4.08, \beta_c = 0°$ (b) $v_1 = 10\frac{m}{s}, \omega = 163\frac{1}{s}, \lambda = 4.08, \beta_c = 0°$

(c) $v_1 = 13\frac{m}{s}, \omega = 196\frac{1}{s}, \lambda = 3.77, \beta_c = 4.88°$ (d) $v_1 = 14\frac{m}{s}, \omega = 196\frac{1}{s}, \lambda = 3.50, \beta_c = 10.00°$

(e) $v_1 = 15\frac{m}{s}, \omega = 196\frac{1}{s}, \lambda = 3.27, \beta_c = 13.37°$ (f) $v_1 = 16\frac{m}{s}, \omega = 196\frac{1}{s}, \lambda = 3.06, \beta_c = 16.08°$

Figure 7.8.: Pressure coefficient on blade sections for VS-VP

8
Experimental Investigations on the Influence of Turbulence

In this chapter the performance of the optimized wind turbine, which has been analyzed in chapter 6, and validated numerically as shown in chapter 7, is experimentally investigated under the influence of the turbulence. A special designed setup with turbulence grids that helps in generating different turbulence levels are presented here. Upwind and downwind velocity distributions were measured to highlight the flow expansion of the three following cases: without grid, with fine grid, and with coarse grid. Furthermore, the tip vortices were analysed for the three cases above.

For individual investigations of the turbulence effect, the tip vortices were isolated with using of a tailored winglet. Hence, additional influences such as, wake-surrounding interaction, delay of boundary layer separation and penetration of oncoming turbulence flow as well as the tip vortices damping were separately investigated.

8.1. Grid-Generated Turbulence

To quantify of the turbulence created by the grids, hot-wire measurements were taken inside the test section. Since the grid turbulence is nearly isotropic [37], the measurements were restricted to the longitudinal direction. Hot-wire and Pitot-tube were

8. Experimental Investigations on the Influence of Turbulence

placed in the center of the flow and traversed in the flow direction during the measurements.

The area near to the grid ($x < 40\ cm$) was excluded from the measurements, since it showed high inhomogeneity of the mean velocity. For this reason, only measurements for $x \geq 40\ cm$ could deliver reasonable results. Steps of $\Delta x = 20\ cm$ were chosen as a compromise between accuracy and expenditure. The results are presented below.

Turbulence Intensity

Turbulence intensity shown in Figure 8.1 stays constant over the whole test section at $TI = 0.007$ when it is measured without the grid. The turbulence level increases with the installation of the fine grid at the entrance of the test section. The range is between $TI_{f,40} = 0.025$ and $TI_{f,160} = 0.011$. The non-linear decrement is caused by the dissipation ε as will be shown in the next section.

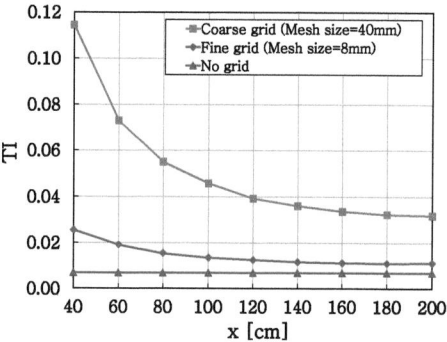

Figure 8.1.: Intensity of grid-generated turbulence along the test section for the three cases (no grid, fine grid, coarse grid)

The same effect, but with a higher level of turbulent intensity, is depicted when using of of the coarse grid. The non-linear range of the turbulence intensity is from $TI_{c,40} = 0.114$ to $TI_{c,180} = 0.032$.

Energy Spectrum

The energy spectrum at different hot-wire distance from the grid (and therefore TI) for various wind velocities are presented in Figure 8.2.

The trends show an increment of $E(f)$ with increasing of distance and decreasing of velocity within the low frequency range. Both facts fit with fig. 8.4 (a) in the next section. The size of the integral length scales is increasing with higher distance and

8. Experimental Investigations on the Influence of Turbulence

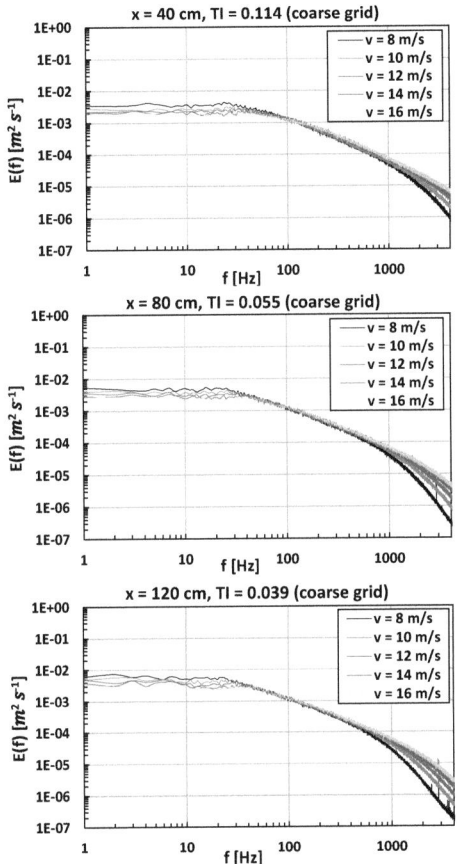

Figure 8.2.: Distribution of Spectra $E(f)$ with eddies frequencies f at different oncoming wind velocities v_{wt} and turbulence levels for changing hot-wire positions x

lower velocity. Since larger eddies contain more energy, this increment is expected. For the high frequency range the tends of $E(f)$ are in opposite. The higher the distance the more the dissipation, this is in agreement with the reduction of TI shown in Figure 8.1. The velocity has the same effect of the distance in reduction of ϵ.

Turbulent Length Scales

The results of the length scale analysis, which were calculated by using equations presented in section 2.6.3, are shown in the following. Figures 8.3 and 8.4 show and compare the Integral length scales (a) and the Taylor micro scales (b) as a function of distance x from the grid and wind velocity v. The ratio λ/L for longitudinal (a) and transverse (b) directions can be seen for different turbulence levels in Figures 8.5 and 8.6. The development of Re_λ as a function of x and v (a) and Re_M and x (b) is shown in Figures 8.7 and 8.8, again for varying turbulence level.

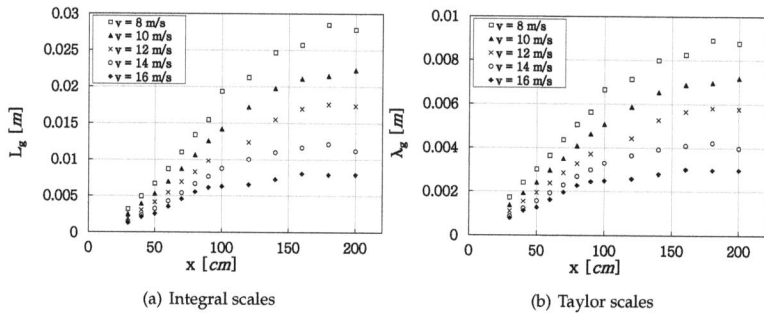

(a) Integral scales   (b) Taylor scales

Figure 8.3.: Length scale distribution at medium turbulence level (fine grid)

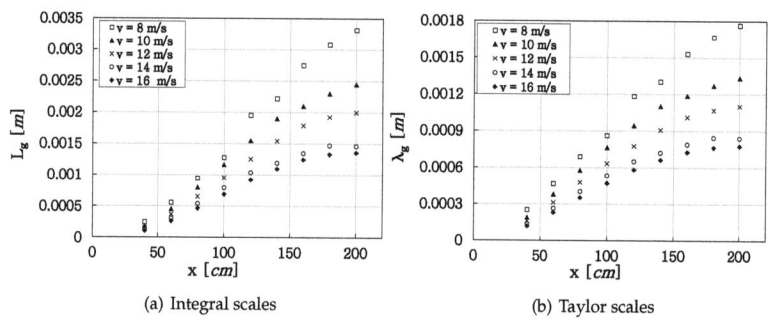

(a) Integral scales   (b) Taylor scales

Figure 8.4.: Length scale distribution at high turbulence level (coarse grid)

8. Experimental Investigations on the Influence of Turbulence

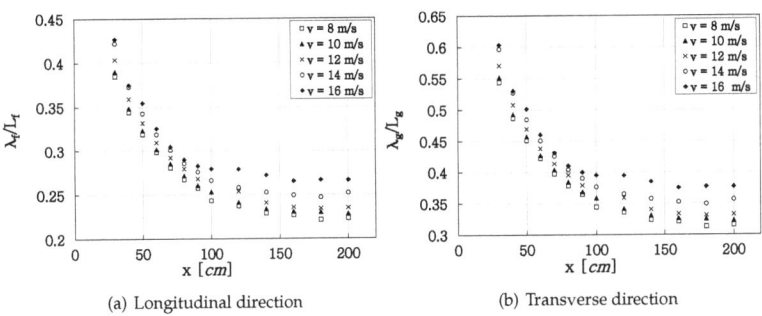

(a) Longitudinal direction

(b) Transverse direction

Figure 8.5.: Ratio λ/L at medium turbulence level (fine grid)

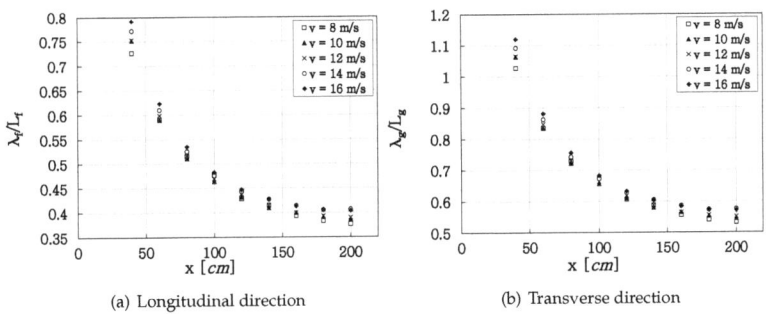

(a) Longitudinal direction

(b) Transverse direction

Figure 8.6.: Ratio λ/L at high turbulence level (coarse grid)

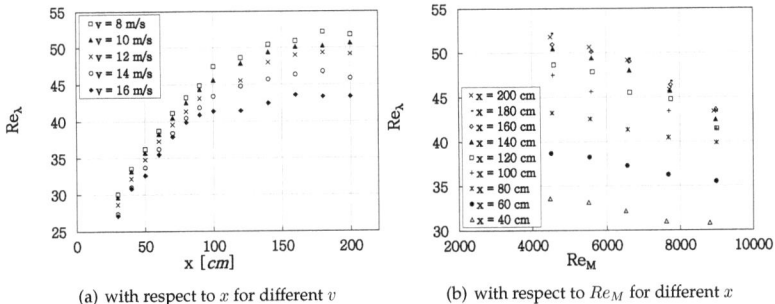

(a) with respect to x for different v

(b) with respect to Re_M for different x

Figure 8.7.: Distribution of Re_λ at medium turbulence level (fine grid)

8. Experimental Investigations on the Influence of Turbulence

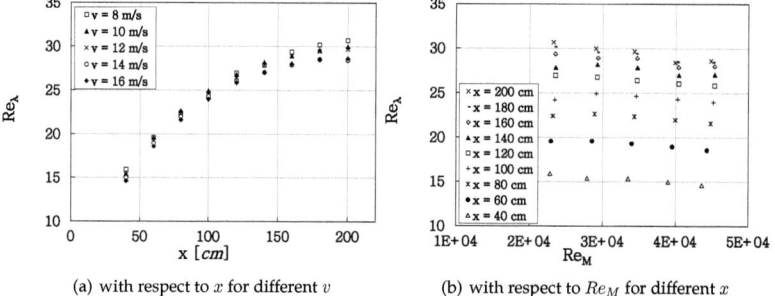

(a) with respect to x for different v

(b) with respect to Re_M for different x

Figure 8.8.: Distribution of Re_λ at high turbulence level (coarse grid)

8.2. Measurements of the Power Coefficient

Twe turbines that designed for high and low Reynolds numbers were consecutively installed in the wind tunnel. They were exposed to the grids with different mesh sizes, hence, two different turbulence levels and varying wind speeds in a range from 10.5 to 13.5 m/s for the $SG6043$ profile and from 14 to 17 m/s for the $S809$ profile. This will guarantee testing the design and the off-design conditions, since the turbines were designed for wind velocities of 12 m/s and 15 m/s, respectively. Both turbines performance will be compared to the their reference performances when operating in the wind tunnel without grids. To study the influence of a wide range of turbulence intensities, the turbines were mounted on a traverse to allow moving them from $x = 60\ cm$ to $x = 120\ cm$ distance from the inlet of the test section in steps of $\Delta x = 20\ cm$, figure 4.2.

8.2.1. Performance Measurements With Fine Grid

Figures 8.9 and 8.10 show the trend of the power coefficient C_P as a function of the tip speed ration λ along the flow direction for both profiles at different positions. For better comparison, C_P is also shown as a function of λ at the free flow turbulence without grid. The power coefficient increases as the turbulence level increases. Moving the turbine towards the grid to face higher TI, the tip speed ratio increases too. This indicates that the rotational speed increases, since the reference oncoming wind velocities are same 8.9. The maximum for each distance is at the design wind speed of $v_1 = 15\ m/s$. This indicates that the stall delays, which is due to increasing of rotational speed for the same wind velocity. For the low Re turbine of $SG6043$ airfoil the maximum values of the power coefficient at all positions is between wind speeds of $12 - 13\ m/s$. The increment of the C_P and λ follow the same trend of the high Re turbine, but they show smoother performance in all positions. Compared to the S809 performance, Figure 8.10 depicts at minimum distance from the grid the maximum performance points are not changing. This is another result beside the smooth trend of the performance, because this turbine is already designed with low Re airfoil, which is within the operation range here. Hence, the turbine is not prone to high stall in its entire blade range.

The turbine reaches high values near to the Betz limit. Hence, an intensive investigations are decided to be done to figure out whether that increment is due to additional energy gaining from turbulence, or additional interactions of turbulence eddies with boundary layer, tip vortices or other reasons that contributes individually or together.

8. Experimental Investigations on the Influence of Turbulence

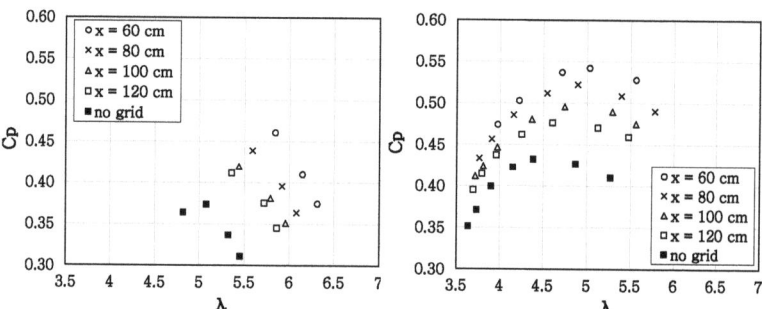

Figure 8.9.: Trend of C_P for the $S809$ profile at varying turbine axial positions with fine grid

Figure 8.10.: Trend of C_P for the $SG6043$ profile at varying turbine axial positions with fine grid

8.2.2. Performance Measurements with Coarse Grid

The high turbulent cases are shown in Figures 8.11 and 8.12. In comparison to the medium turbulence levels, high turbulence shows a higher performance. Thus, performance curves are shifted to the right for both turbines due to higher rotational speeds.

Figure 8.11 shows that the maximum performance locations of the high Re turbine are moved from wind velocities of $v_1 = 15\ m/s$ to $v_1 = 14\ m/s$. Here, the blockage of the turbine becomes too high because of the higher rotational speed (higher λ). The flow cannot any longer pass through the rotating blades as easy as before. The increment on λ is an indication for a higher rotational speed compared to the medium turbulent case before, since the wind velocity is the same. The non-stall regions extend for all positions to reach velocities less than the design velocity of $v_1 = 15\ m/s$, which indicates that the more power extraction from high turbulence contributes in enhancing the performance with more reduction of the stall. The low Reynold no. turbine with its thinner $SG6043$ profile is not influenced by this effect. Here, the increment is even higher. The power coefficient can reach values, which is even closer to Betz limit. Trying to explain this effect, detailed experiments are discussed in the following sections.

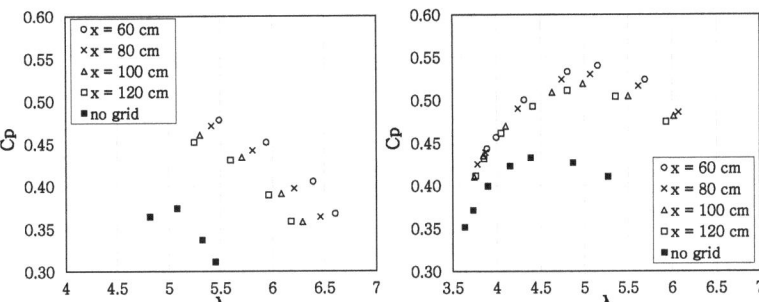

Figure 8.11.: Trend of C_P for the $S809$ profile at varying turbine axial positions with coarse grid

Figure 8.12.: Trend of C_P for the $SG6043$ profile at varying turbine axial positions with coarse grid

8.3. Velocity Distribution

It has been shown in section 8.2 that there is an increment in C_P, especially for the SG6043 profile. In order to understand this effect, the velocity distribution around the turbine was measured during the operation of the turbine to provide insight of the upwind velocity deficit. The wind turbine is set on a wind tunnel test section at a sufficient distance, which is chosen to obtain the reference oncoming wind speed. The first indication for the accurate reference wind velocity is when its distribution across the far upwind lateral sections (or section of symmetry) is constant, which implies there is no wind turbine blockage effect. Second indication is when the velocity in the axial direction towards the turbine stays constant for a sufficient distance before it then drops. This implies that the axial position of the turbine in relation to the wind tunnel is sufficient to appoint the wind tunnel velocity as a reference. Both requirements are fulfilled at a turbine distance of $x = 120\ cm$ from the inlet of the test section.

8.3.1. Velocity Distribution at Free Flow Turbulence (Without Grid)

In Figure 8.13 it is shown that at a dimensionless ratios of the upwind distance to rotor diameter of $x/D = -1.7$, the lateral y/D distribution of velocity is constant. It will keep constant values in both axial and lateral distributions until reaching the dimensionless upwind distance of $x/D = -1.5$.

In Figure 8.13 also, as we move towards the turbine, i.e. increasing x/D, the velocity distribution drops and the dropping area becomes wider and wider. This drop is associated with the existence of the turbine, where it extracts the incoming wind kinetic

8. Experimental Investigations on the Influence of Turbulence

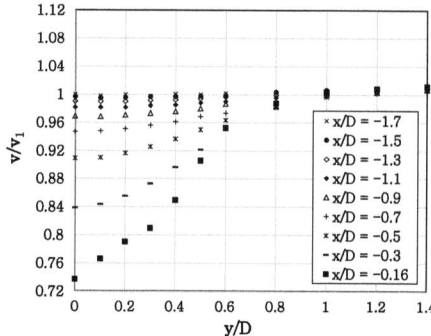

Figure 8.13.: Upwind velocity distribution without grid

energy and converts it to a mechanical one as well as the rotor stagnation Thus, the velocity drops to a value which in turn indicates the turbine performance C_P (equation 2.11). The dropping is reduced as we move in the lateral direction and it even exceeds the turbine radius at $y/D = 0.5$. It is clear in the figure the distance where the velocity deficient is still obvious until $y/D = 1$.

Figure 8.14(a) shows the velocity distribution behind the turbine. Moving away from the center in lateral direction, there is a decrement of velocity. This indicates that the extracted energy is increasing as we move outboard and the maximum extraction appears at $y/D = 0.4$ or 80% of the blade, where the maximum velocity deficit. At the near weak region $x/D = 0.2$, starting from $y/D \approx 0.5$, which is exactly the radius of the turbine, an abrupt increment can be seen clearly. Here is the border between the areas of the extracted wind kinetic energy and the surrounding unaffected flow, which represents the surrounding stream tube. For a better view, this area is shown enlarged in fig. 8.14(b). There can be seen an increase in velocity of about 20% compared to the reference (12 m/s, or 1 when normalized to it self). Since the turbine is blocking the flow, the surrounding flow just around the rotor bends around the rotor tips results in a higher flow rate which leads to an increment of the velocity at that region. The region between the edges of the abrupt increase and the points before the increment is the boarder of the stream tube, within it the tip vortices have their influence. This region is expanding in the wake and still refer to the wake shear inner and outer boundaries, where in which the shear resulting from different velocities is appearing. Hence, as we move away from the turbine in axial direction x/D, the influence of the blockage becomes less, and hence, the mixing area becomes wider. In fig. 8.15, the approximate development of the stream tube is plotted with the dashed curves. The velocity deficiency will extend more outside and the drop is becoming more and more flat. It can

8. Experimental Investigations on the Influence of Turbulence

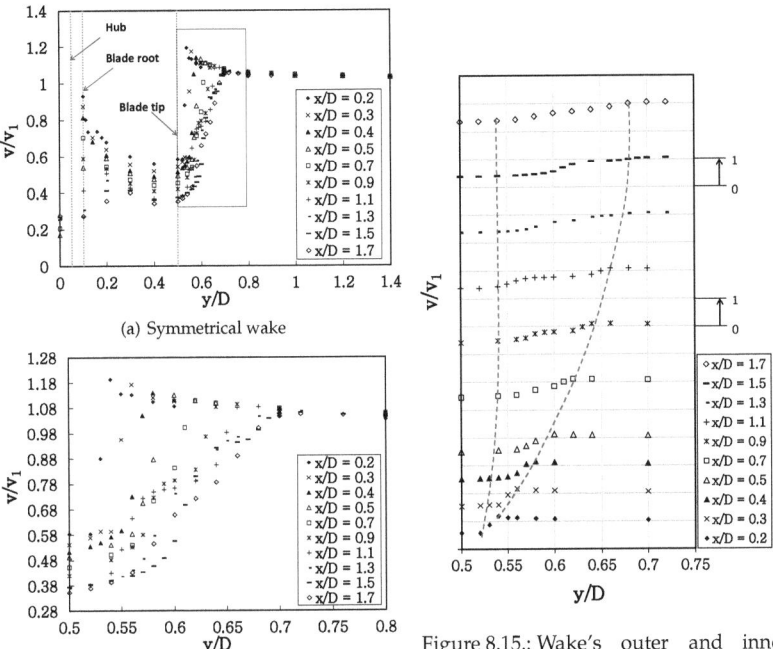

Figure 8.14.: Downwind velocity distribution without grid

(a) Symmetrical wake

(b) Wake shear border

Figure 8.15.: Wake's outer and inner symmetrical borders without grid

be also seen that at higher distances from the rotor, velocity in the area of the turbine shrinks. This fact agrees with the one-dimensional theory, where the average axial velocity keep reducing from upwind v_1 to downwind v_3 passing through the turbine v. The flatness and the smoothness in the velocity distribution are indications for the increasing of the mixing area between the wake and the surrounding, which indeed increases as we move downwind where the low pressure wake entrains more energy from surrounding. Furthermore, in Figure 8.14(a) there is a local maximum in wind speed at about $y/D = 0.1$ and $x/D = 0.2$ and it expands more and more until it located at $y/D = 0.3$ and $x/D = 1.7$. Here is the point of the blades intersection with the generator or the blades roots, where the oncoming velocity penetrates through. Thus, minimum reduction in velocity appears and expands in lateral direction as we move in axial direction due to the expansion in the wake area. Due to the dimensions of the generator, it was not possible to take any measurements along the axis of rotation

8. Experimental Investigations on the Influence of Turbulence

$x/D = 0$ directly after the turbine.

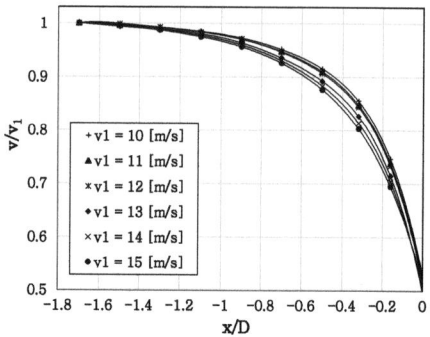

Figure 8.16.: Upwind velocity distribution along the axis of rotation ($y/D = 0$) without grid

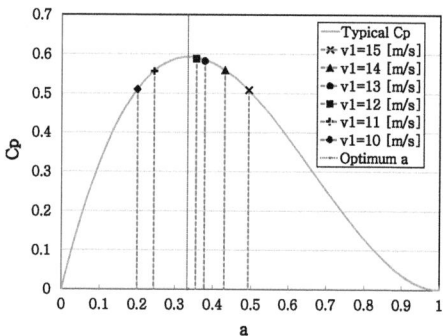

Figure 8.17.: Typical power coefficient curve and the operating points at different wind velocities

All measurements until this point were taken at $v_1 = 12\ m/s$, which is the design velocity of the low Reynolds turbine, since the aim is to study the design performance at free flow turbulence and compare it later with higher turbulence levels. Thus, velocity distribution at different wind speeds is less important. The only important velocity distribution is the axial upwind velocity distribution, which is needed to prove that at the design velocity the turbine performance is maximum. In Figure 8.16, data are taken at velocities starting from 10 to m/s. The maximum wind speed was limited by the experimental setup, where above 15 m/s, vibrations became too strong. It can be seen that the extend of influence of the turbine on the wind speed is proportional to the

8. Experimental Investigations on the Influence of Turbulence

reference velocity, starting from $v_1 = 1.25\ m/s$. This is reasonable, the higher the reference, the more the kinetic energy can be extracted from the wind. This does not mean that C_P increases too. Since the power coefficient is defined as the ratio of the extracted kinetic energy to the oncoming wind energy, thus, highest efficiency of the turbine is expected at the steepest drop of velocity, which is indicated by the axial induction factor a, equation (2.11). Due to the dimensions of the Pitot-tube, the wind velocity could not be measured at distances less than $x/D = 0.16$, so a trend-line was added. It is chosen as polynomial trend-line of fourth order, because it fits the measured data.

By using the polynomial equations to extrapolate the velocity at the turbine section ($x/D = 0$), it is possible to estimate the axial induction factor a for each reference wind velocity v_1, equation (2.6). Equation (2.11) was used to predict the typical C_p, in which the optimum value of a is $1/3$. Then, the estimated values of a are projected to the C_p curve to predict the values of C_p for different operating points, as shown in Figure 8.17. The figure shows that the maximum C_p is obtained at a reference wind speed of $12\ m/s$, which is exactly the turbine design velocity. This investigation procedure can perform only for a turbine of the same turbulence level and to compare within different velocities. It cannot be generalized to compare for example a defined turbulence level with others of higher turbulence levels, since many other parameters could appear that influence the down as well as the upwind, hence, we need to follow the reduction in the axial velocity until far wake distance beside other parameters (they will be explained in details in later sections), which indicate the highest energy extraction.

8.3.2. Velocity Distribution at Medium Turbulence Level (With Fine Grid)

Generally, the velocity distribution at medium turbulence level shows the same behaviour the free flow turbulence.

The upwind velocity distribution at medium turbulence level is shown in Figure 8.18. In comparison to the free flow turbulence, there is a cognizable increment in velocity of a few percent at radius higher than $y/D = 1$. This is due to the higher contraction of the flow at the center of the wind tunnel test section (or the position of the wind turbine) when using of grids. Hence, a smaller cross sectional area and a higher wind velocity to satisfy the continuity.

There is no difference of influence by the turbine in far distances more than $x/D = 1$. The velocity drops not significantly more than in the case of no grid. More energy is extracted from the wind, which results in a higher C_P, but this can not only refer to the extraction of kinetic energy in front of the turbine. Of course, there could be a significant higher drop at distances less than $x/D = 0.16$, but this could not be investigated.

8. Experimental Investigations on the Influence of Turbulence

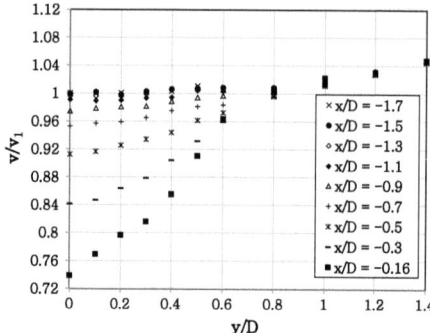

Figure 8.18.: Upwind velocity distribution with fine grid

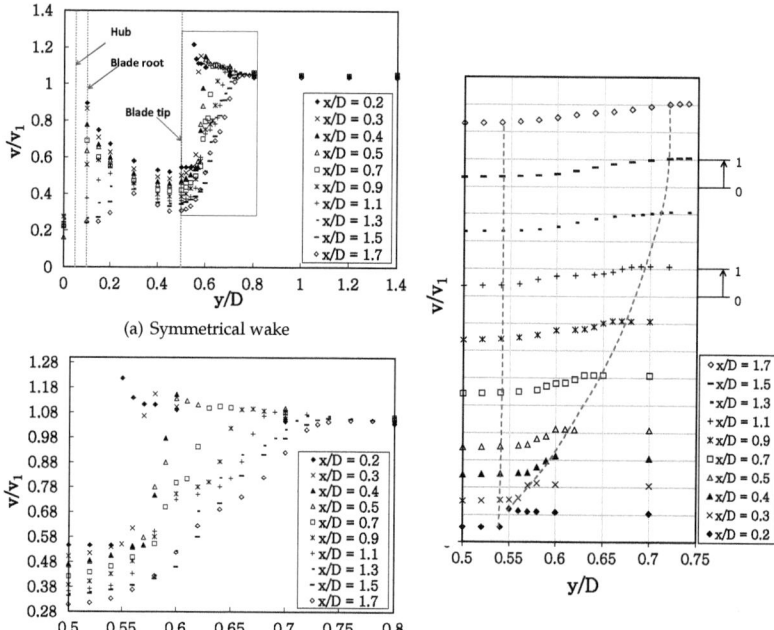

(a) Symmetrical wake

(b) Wake shear border

Figure 8.19.: Downwind velocity distribution with fine grid

Figure 8.20.: Wake's outer and inner symmetrical borders with fine grid

If we take a closer look on Figures 8.19(a), 8.19(b) and 8.20, which show the velocity

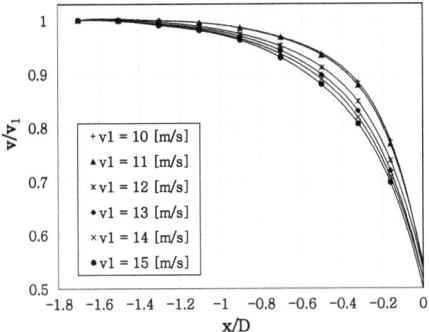

Figure 8.21.: Upwind velocity distribution along the axis of rotation with fine grid

distribution behind the turbine at medium turbulence level, it becomes apparent that the increment of velocity at the tip vortex area of the rotor is wider than in the free flow turbulence case. This in turns refer to increment in the energy extraction. Furthermore, the drop in velocity in the axial downwind direction is another indication for the higher energy extracted by the turbine. Figure 8.21 shows the upwind drop of the oncoming velocities for the medium turbulence case. the decrement is highest at the design wind velocity, this can be shown as has been done to the free flow turbulence (no grid case), Figure 8.17.

8.3.3. Velocity Distribution at High Turbulence Level (With Coarse Grid)

As in the medium levels, the higher turbulence level will only differ in the intensity of the effects on the velocity distribution. Figure 8.22 shows the drop in the upwind velocity.

The downwind distribution of velocity in Figures 8.23(a), 8.23(b) and 8.24 show more flatness and smoothness than low and medium turbulent levels. In the wake of the rotor there is no significant difference, but the mixing tip vortex border of the surrounding stream tube is becoming wider. At a very near distance of $x/D = 0.2$, the mixing is already starting at $y/D = 0.54$ and ends first at $y/D = 0.8$. Even at far distances, the mixing area continues to stay wider. The comparison of fig. 8.23 with Figures 8.14 and 8.19 shows this very clearly.

The upwind drop of the axial velocity is shown in figure 8.25. Again by following the same procedure applied to get Figure 8.17 is applied to show that the maximum

8. Experimental Investigations on the Influence of Turbulence

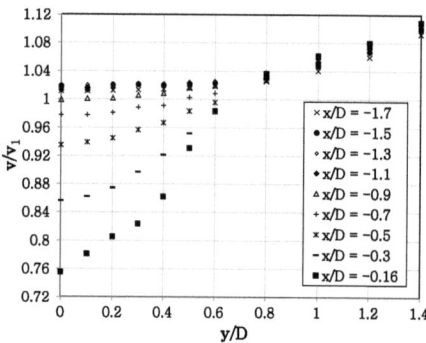

Figure 8.22.: Upwind velocity distribution with coarse grid

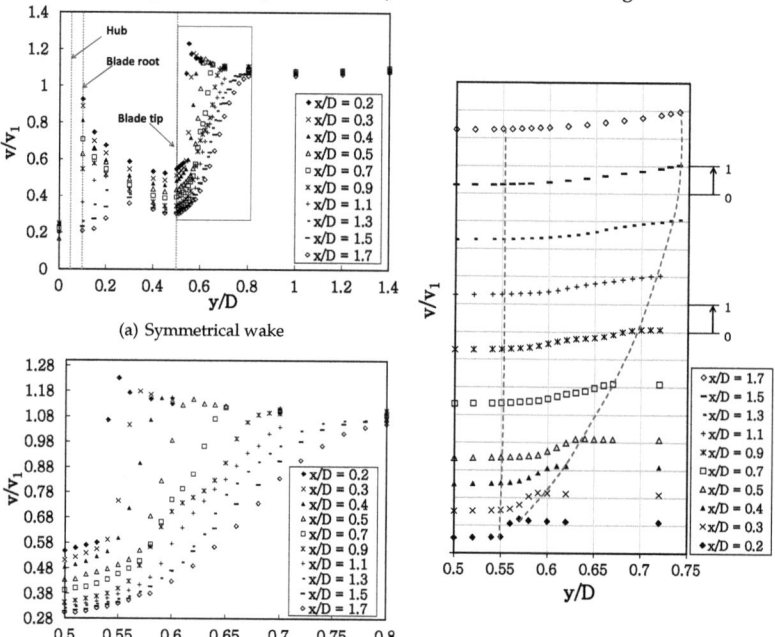

(a) Symmetrical wake

(b) Wake shear border

Figure 8.23.: Downwind velocity distribution with coarse grid

Figure 8.24.: Wake's outer and inner symmetrical borders with coarse grid

8. Experimental Investigations on the Influence of Turbulence

C_P is at $v_1 = 12 \ m/s$, which is identical with previous results of the no-grid and the fine-grid.

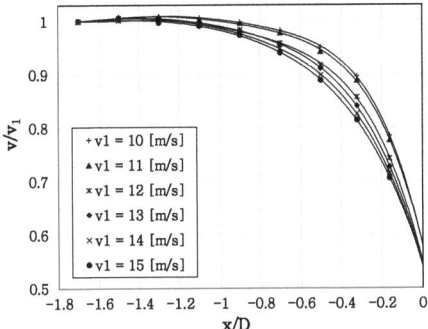

Figure 8.25.: Upwind velocity distribution along the axis of rotation with coarse grid

As a result, velocity distribution investigations show not much difference in the upwind velocity for different turbulence levels. whereas the downwind velocity distribution shows clear evidence for C_p increment. Hence, it is expected that there is no direct contribution of turbulence on the oncoming wind kinetic energy but rather its fluctuations could interact with the rotor blades and this interaction is the reason of the enhancement of C_p. The interaction could be with the boundary layer, tip vortex and penetrating of turbulence within the wake either through the rotor or entrainment from the downwind wake surrounding. Thus, the need for individual investigations is mandatory to reveal the dominant influence.

8.4. Influence of Turbulence on Tip Vortex

Recall, tip vortex is one possible expected explanation of what was found in the influence of the power extracted. It is clear from the downwind investigations that they are influenced by the turbulent levels. Not only the flow expansion with highest power extraction is effected, but also the border of the wake stream tube. For this reason, additional measurements of velocity were taken in an area behind the tips from axial distances between $x/D = 0.2$ to $x/D = 0.6$ in steps of $\triangle x/D = 0.2$. Due to the highest influence is in the very near weak, only measurements to $x/D = 0.6$ will be analyzed here. The maximum resolution in radial direction was $\triangle y/D = 0.01$, which is equal to 5 mm and the measurement time is 1 $second$, reference velocity was set to $v_1 = 12 \ m/s$. For each x/D, the radial direction with the highest fluctuation in velocity

8. Experimental Investigations on the Influence of Turbulence

was chosen and directly compared to the corresponding y/D for the other turbulence levels. Highest fluctuation represents the maximum influence of the tips.

8.4.1. Tip Analysis

At the axial distance of $x/D = 0.2$ and the radial distance of $y/D = 0.52$, as shown in Figure 8.26 (a). This is the position of the maximum fluctuation for free flow turbulence. Here, the influence at medium and high turbulence level is not that high yet. By trend, with increasing turbulence intensity, the downstream velocity is decreasing. This fact agrees with the power coefficient measurements. Measured velocities are less than the reference velocity of $v_1 = 12\ m/s$ for all cases, because the measurements were taken inside the wake of the turbine. There is a second kind of fluctuation recognizable, it can be clearly seen for the coarse grid case, which may come from the high turbulent scale eddies.

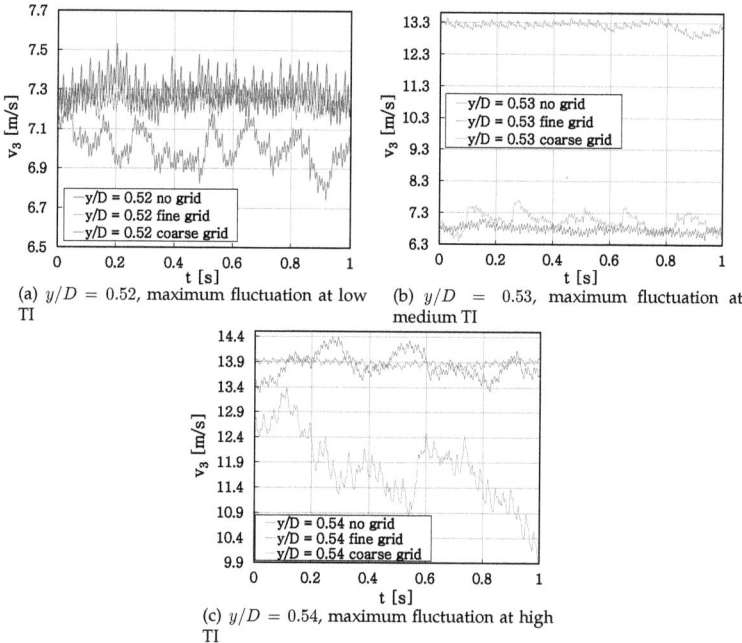

Figure 8.26.: Tip vortices at $x/D = 0.2$ for different turbulence levels

As can be seen in Figure 8.26 (b), by a movement of $\triangle y/D = 0.01$, there is a enor-

8. Experimental Investigations on the Influence of Turbulence

mous jump in velocity (c.f. fig. 8.14). At this position, it is outside the wake for free flow turbulence, but still inside for medium and high turbulence levels. This presents in consistense with the previous results, the medium and high turbulent wakes are wider than the low turbulent wake. There is a different in the tip humps values, this is because the pitch angle for one blade is just a bit different from that of the other blades. Counting 32 maximums within time period of $t = 1$ s leads to a angular speed of $\omega \approx 200$ rad/s, which indicates that these humps is directly related to the rotor tips. At this position, the fluctuation is maximum for the fine grid case.

At a radial distance of $y/D = 0.54$, there is maximum fluctuation for the high turbulence case with the coarse grid. The result is shown in 8.26 (c). Here, it is not exactly clear, if the high turbulence measurement was taken inside or outside of the wake. It is possible that it was somewhere in between, but for low and medium turbulence cases it is very clear. Again, here is still the influence of the tips. While with the use of the fine grid, there is an explicit influence of the mixing of the eddies and there is no influence at all for the no turbulence case.

At the axial position $x/D = 0.4$, the radial position with maximum fluctuation for the low turbulence measurement is at $y/D = 0.56$ as shown in fig. 8.27 (a). In comparison to the $x/D = 0.2$, the steam tube is wider. The absolute fluctuation is becoming a bit less and there is more mixing inside the wake, since there are different eddy scales mixing with the tip vortices. This is the case but more obvious for fine and coarse grid. Here, the pitot tube was again placed inside the wake.

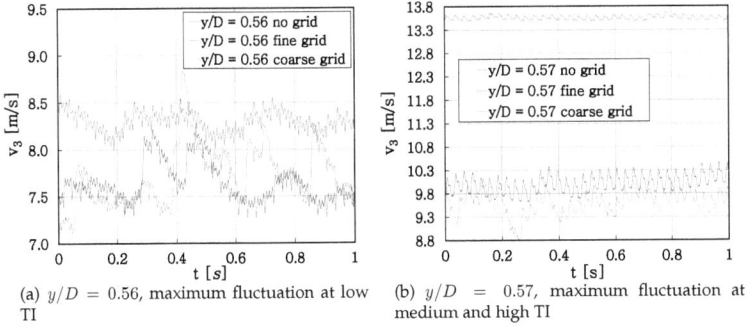

(a) $y/D = 0.56$, maximum fluctuation at low TI

(b) $y/D = 0.57$, maximum fluctuation at medium and high TI

Figure 8.27.: Tip vortices at $x/D = 0.4$ for different turbulence levels

In contrast, for a distance of $y/D = 0.57$, the low turbulent vortices is outside the wake, but not for the medium and high turbulent flows, Figure 8.27 (b). The number of tips maximums can be counted here, about 33, this means the angular speed of $\omega \approx 206$ rad s^{-1} which is higher than without grid. For the high turbulent case, the number

8. Experimental Investigations on the Influence of Turbulence

is even higher. This indicates an increase of the angular speed with the increment of turbulent levels, as concluded previously. It becomes clear in Figure 8.28 (a), that

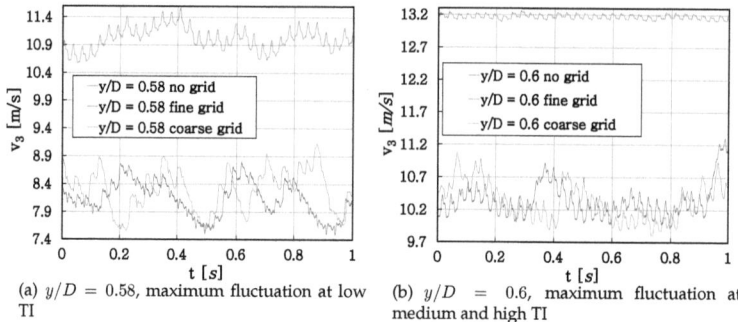

(a) $y/D = 0.58$, maximum fluctuation at low TI

(b) $y/D = 0.6$, maximum fluctuation at medium and high TI

Figure 8.28.: Tip vortices at $x/D = 0.6$ for different turbulence levels

the mixing area is becoming wider and wider. The velocity in the low turbulent case drops to an average of approximately 11 m/s. Even for the fine and coarse grids, a huge influence of the mixing with surrounding wind is obvious. The gap between the expansion of the stream tube becomes wider with higher x/D distances from the rotor plane. Moving with $\Delta y/D = 0.02$ outside (fig. 8.28 (b)), the maximum fluctuation for the fine grid use is the same as the coarse grid use. The difference between both in mixing and velocity appears more and more.

Offering a more detailed view, the tip vortex is visualized for one single revolution of the rotor in the following Figure 8.29. The one revolution tip vortex is obtained by splitting the tip vortices shown in previous figures (8.26-8.28) into correspondence angular speed ω of each, and hence f values ($\omega = 2\pi f$). Then the set of data were averaged to obtain only one rotational averaged vortices. The velocity is normalized by its mean value to simplify the comparison. For a full revolution, three tips are expected to be seen. Each minimum appears when a rotor blade is passing the Pitot tube. It can be seen very clear that not only the tip vortices are suppressed (less amplitude) with increasing of turbulence, but also, the revolution speed increases (less time). Simultaneously, the mixing with the surrounding increases. Figure 8.29(b) shows that vortices are damped for the case of fine grid, in spite of that, the humps of the vortex are still appearing. Whereas for the coarse grid case 8.29(c), the additional damping delays until the distance of 0.4, where there is no clear vortex humps. There is only a higher mixing due to larger eddy scales containing in the oncoming turbulence.

8. Experimental Investigations on the Influence of Turbulence

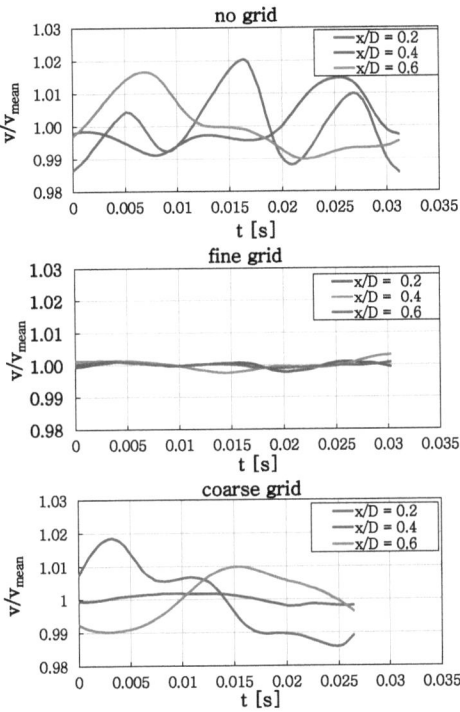

Figure 8.29.: Comparison of the tip vortex for a single revolution of the rotor at varying x/D and turbulence level

8.5. The Effect of Winglets

Previous investigations have shown an effect of turbulence on the power coefficient. More intensive investigations in the tip vortex region reveal the increase in mixing with the increment of turbulent level. Precise tip analysis have shown an expansion of the wake stream tube downwind the turbine. There also was an indication of a higher rotational speed caused by the turbulence. In order to isolate the influence of tip vortex from other factors that possible contributes in the performance increment, a winglet was added to the tip.

The winglet shown in Figure 8.30 was added to the turbine rotor tips and downwind velocity distribution measurements were repeated. Since the influence of winglets is restricted on the downwind area, upwind experiments were not executed. Preliminary

8. Experimental Investigations on the Influence of Turbulence

measurements show that the effect of winglet is not the same for all turbulence levels since the vortices are in different levels.

Results for higher turbulent levels show no influence in C_P when adding winglet. This is because the tip vortices have already damped by the turbulence as has been shown in Figure 8.29. Therefore, investigation of winglet was limited to the case of low turbulent level to compare it later with higher turbulent levels. Winglets supposed to suppress the tip losses and result in higher power extraction due to higher extraction of kinetic energy. Thus different designs were tested until reaching a noticeable power increment.

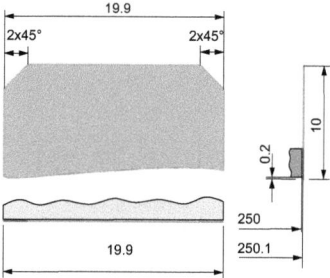

Figure 8.30.: Winglet design employed in the experiments. Dimensions are in mm

Figure 8.31 shows the increment of C_p by the use of the winglets from 8.30 compared to the measurement without winglets at free flow turbulence (no grid). The gain is noticeable very clearly, but still the effect is less than the influence of adding turbulence when using of both fine and coarse grids as shown in Figure 8.31. This is a proof for the expectation, that turbulence not only helps in suppressing the tip losses, but there are additional effects, such as boundary layer interaction, stall . For an easier comparison, fig. 8.31 (b) shows the relative increment based on the no winglet use. Here it can be seen that the increment is on average about 9 % for a range of $\lambda = 4.2 \ldots 5.6$, which represents oncoming wind velocities from about $v_1 = 10 \ m/s$ to $v_1 = 15 \ m/s$. Outside this interval, the effect declines. Since the winglets do not have any influence on the upstream flow, further investigations are constricted to the downwind.

As can be seen in Figure 8.32 (a), the winglet have an influence on expanding the wake as compared to no winglet turbine wake. Figure 8.32 (b) offers an enlarged view on the corresponding area. In direct comparison to 8.14 (b), there is a shift of the velocity mixing area to higher y/D at the inner, as well as the outer border. Since this shift is not too distinctive, it is easier to compare Figures 8.33, 8.15 and 8.20. Wake comparisons show that with winglet has wake size larger than no grid case and smaller than fine grid case. This indicates, that the winglets are suppressing the tip influence.

8. Experimental Investigations on the Influence of Turbulence

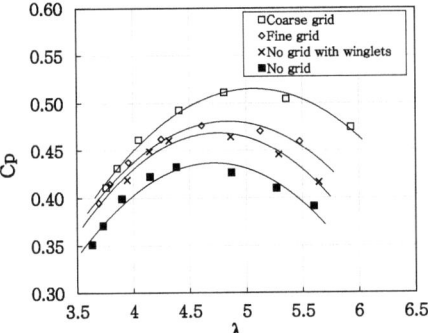

Figure 8.31.: The effect of winglet on C_P at free flow turbulence (without grid). All cases at wind turbine position of x=120cm

In comparison to Figures 8.20 and 8.24, they both show wider wake and hence more power extracted. This implies suppressing the tip vortex is not the only reason for the increment of C_P with the increase of turbulence levels. Hence, it was planed doing investigations of spectral energy to show more details of the wake-surrounding interactions in both stream tube boundary and the possible turbulent penetrates through the rotor

8. Experimental Investigations on the Influence of Turbulence

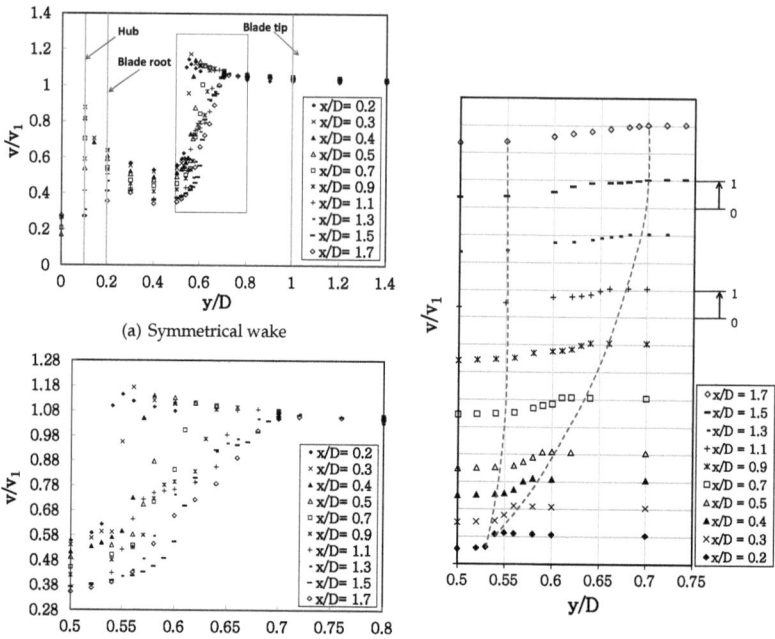

Figure 8.32.: Downwind velocity distribution without grid and with winglets

(a) Symmetrical wake

(b) Wake shear border

Figure 8.33.: Wake's outer and inner symmetrical borders without grid and with winglets

8.6. Turbulence Measurements

To finally clue the puzzle about the C_P increment, hot-wire measurement were used to measure the turbulence intensity and energy spectrum during turbine operation. The upwind distribution of TI and the scales are already known from chapter 8.1, so measurements were only conducted in downwind wake. Measurements of turbulence intensity and spectrum energy can highlight more details of how energy exchanges between the surrounding and the wake and more intensive investigations in the interior wake region will show the penetrating of the oncoming flow through the rotor blades.

In order to follow the wake development, measurements of TI were taken at the five downwind axial positions $x/D = 0.3, 0.5, 0.7, 1.1$ and 1.7. For each x/D, a radial scan was conducted within the range of $y/D = 0.2 - 0.8$ which covers the wake from

8. Experimental Investigations on the Influence of Turbulence

an interior point (10 cm above the axis of rotation), passing through the wake shear border until the surrounding undisturbed flow. One measurement was taken at the maximum fluctuation position, which changes for each case. Figure 8.34 compares TI inside the wake for a fixed turbine position of $x = 100\ cm$ and different turbulence levels.

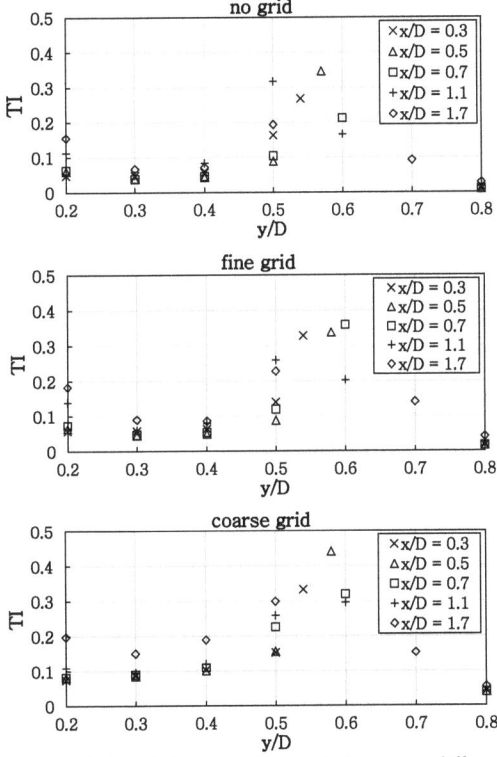

Figure 8.34.: Downwind turbulent intensity distribution at different oncoming turbulence levels for the wind turbine position of 100 cm

It shows that at the interior wake range of $y/D = 0.2 - 0.4$, turbulence penetrating through the rotor increases with higher oncoming turbulence. It also indicates the faster rotational speed of the turbine. In addition, the figure shows an increase in the far wake with higher oncoming turbulence level, which means more wake mixing with the surrounding. It can be seen more clearly mixing and hence adding more turbulent at the shear border region, which is different and expanding axially with the distance.

8. Experimental Investigations on the Influence of Turbulence

It lies for all position in a range of $y/D = 0.5 - 0.6$. It is ought to mention here, that the measurement positions within this region were selected for the maximum tip vortex by using the previous tip vortex information. Again, the high turbulent case shows a wider TI range due to the expansion of the wake shear border. Moving more outside the wake to $y/D = 0.7 - 0.8$, it becomes clear the TI increases, which is indeed expected as it indicates the oncoming wind turbulent level (see section 8.1).

8.7. Spectrum Analysis

For better understanding about the interaction between wake and surrounding and how turbulent energy develops within different scales, the same measured points of turbulent intensities in section 8.6 were used to measure the energy spectrum. Starting with the near wake distance of $x/D = 0.3$, Figure 8.35 shows the case when the turbine is mounted at a test section position of 100 cm from the inlet of the test section, where the oncoming turbulence intensities are known from chapter 8.1 for different turbulent levels. It is possible to distinguish between energy developments as a function of eddies frequencies f at a different oncoming TI. In general, the energy of the most containing frequencies ($f \leq 100\ Hz$) increases. In analogy to the analysis of TI in the previous chapter 8.6, this increment is associated with the penetration of the oncoming turbulence through the rotor plane. Figure 8.35 also shows three distinguished jumps of the energy at defined frequencies ($f \approx 34, 66$ and $100\ Hz$). These peaks correspond to the three rotor blades and represent the additional increment of TI (and thereby turbulent energy) of the rotation of the turbine, which was also mentioned in chapter 8.6 before. As expected, the amplitude decays with increasing distance from the tip at y/D in both directions. Here, the mixing is most distinctive. With additional oncoming grid-generated turbulence, the turbine rotates faster. This fact becomes clear in the figure by the shift of the peaks to higher frequencies. Especially at the maximum fluctuation position (yellow color) for the low turbulent case, there is a high intensive mixing over a wide interval of frequencies. The effect fades with higher oncoming turbulence, for the high turbulent case it is nearly completely disappeared. This is a confirmation of the results in chapter 8.4.1. The fading is also noticeable for increasing measuring distance from the turbine (8.36). Here, the three typical peaks still exist, but tip vortices are vanished through dissipation processes.

8. Experimental Investigations on the Influence of Turbulence

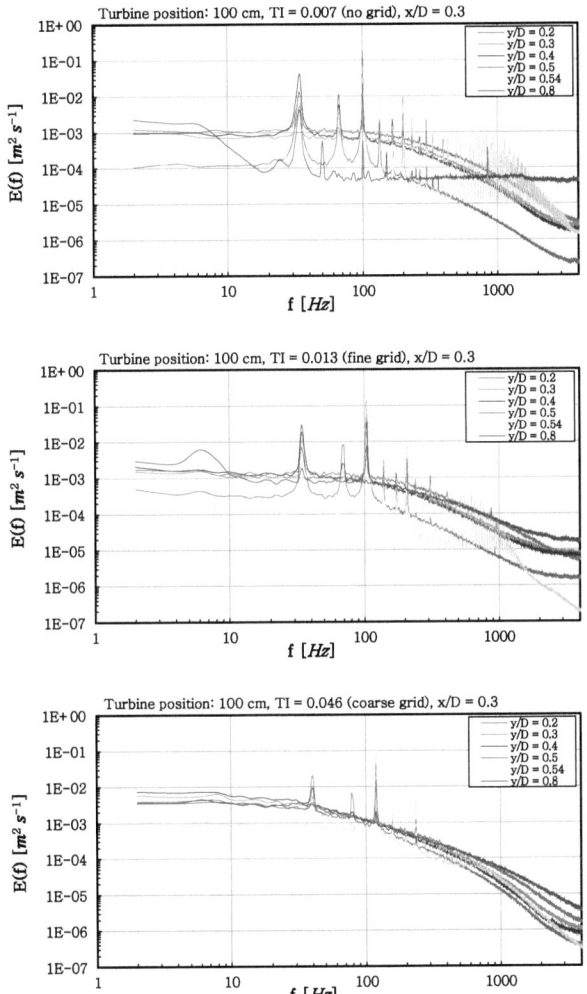

Figure 8.35.: Distribution of Spectra $E(f)$ with eddies frequencies f at different radial distance y/D in the turbine wake for different oncoming upwind turbulence levels at hot-wire downwind position of $x/D = 0.3$

8. Experimental Investigations on the Influence of Turbulence

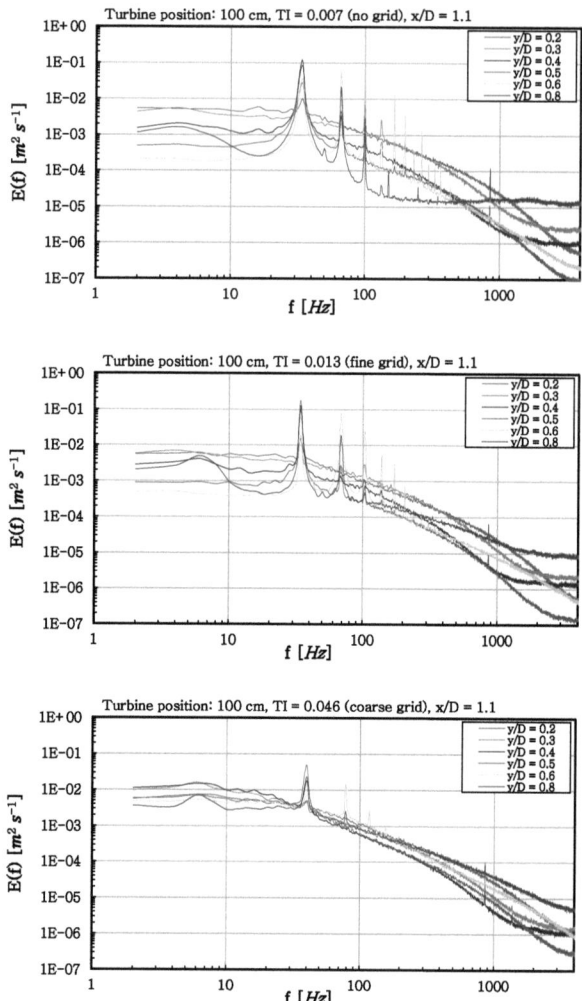

Figure 8.36.: Distribution of Spectra $E(f)$ with eddies frequencies f at different radial distance y/D in the turbine wake for different oncoming upwind turbulence levels at hot-wire downwind position of $x/D = 1.1$

9
Conclusions and Outlook

9.1. Conclusions

The present thesis aims at establishing strategies for investigating and improving the Horizontal Axis Wind Turbines (HAWT) operating under the turbulence conditions. Hence, figuring out the impact of turbulence scales on the performance of the HAWT. This is done by exposing a produced HAWT model to turbulence with various energy and length scale content, which is generated by using two static squared grids. In the following, the current developed tools and methodologies that are used to reach the objective of the present study as well as the concluding remarks of employing them are presented.

For a systematic investigation of the turbulence influence on the wind turbine performance, a laboratory-scale HAWT that efficiently perform in a wind tunnel has been designed. It's efficiency is in the range of the real-scale HAWT. Thus, the obtained measurements are reliable and possible to be reflected and evaluated for different scales of HAWTs. Hence, it is suggested firstly at developing the blade shape design and optimization method called Torque-Matched Aerodynamic Shape Optimization (TMASO) that maximizes the extracted power while constrained to the torque-rotational speed characteristic of the coupled generator.

9. Conclusions and Outlook

The TMASO involves a combination of Schmitz and BEM theories for an initial design of a HAWT which has the advantage of providing a new profile that includes the rotational wake. The interfacing with the XFOIL while optimization proceeding to calculate the lift and drag coefficients of the blade sections at a given speed and angle of attack was very fruitful and deliver a trusted information to the optimization. The torque-matched technique employed in TMASO has proven its superior performance in producing an efficient HAWT model and as alternative technology in designing of HAWTs.

The work involves additional optimization method called Torque-Matched Pitch Control Optimization (TMPCO), which is developed to verify the ability of handling different control-strategies. It is constrained to the torque of the coupled generator and aimed to keep the rated power constant by adjusting the blade pitch angle. Results show that by reducing the angle of attack it is possible to achieve the objective.

In order to predict the performance of the optimized turbine under various operation velocities, an analysis method of the Torque-Matched Aerodynamic Performance Analysis Method (TMAPAM) has been developed. It is shown to monitor the operation performance of an existing HAWT over a selected range of incoming wind velocity while matching the torque generated by the wind turbine to that of the drive unit at a certain rotational speed. The Schmitz and BEM theories were combined in the analysis to reduce the computational effort, so that TMAPAM becomes appropriate for very high number of optimization iterations. The study on the influence of profile and tip losses demonstrated that they have to be included in the analysis to have more realistic performance results.

TMASO is used to produce a laboratory scale wind turbine for experimental tests in the closed loop wind tunnel. Reynold number consideration when down scaling of wind turbine is proven to be essential to keep the performance of the model in the range of the large scale wind turbines. The rotor blades are specially designed and optimized for this wind tunnel and the generator used. A special designed setup that can measure the performance in the wind tunnel has been developed. The generator torque needed for calculation is measured separately with a second experimental setup. The comparison of the experimental results with that of the analysis method showed a good agreement. Although, the proposed TMAPAM is cheaper and faster in comparison to the conventional methods such as full scale simulations and field experiments, it had to be validated for large scale turbines.

9. Conclusions and Outlook

In order to verify the applicability of the developed optimization methods for a real scales HAWTs, they are applied to the NREL stall-regulated 10kW wind turbine. The study showed that TMASO performs well for a large scales wind turbines. The limitation of the chord length is necessary to prevent it from a rapid increase when optimizing of the rotor shape. This developed method serve in changing the rotor shape while keeping its angular speed constant.

The TMAPAM is further developed to predict the performance of the stall-regulated wind turbine. It showed an agreement with the available experimental data, hence, it is proven to be reliable for further investigations. However, for a general performance prediction of the stall-regulated wind turbine, it needs a method for predicting the lift and drag coefficients in addition to consideration of turbine characteristics and operational conditions. The developed performance prediction method, which includes the change in radius and chord length, can predict the performance for a relatively high range of wind turbines scales, number of blades and chord lengths. The post-stall model parameters have to be adjusted before applying it to any wind turbine.

In order to reveal the applicability of the TMPCO method for different wind turbines scales, it is applied to a pitch-controlled NREL 5MW wind turbine. Results show that it can indeed predict the performance with satisfactory agreement with the experimental data. In general, the developed methods provide complete tools for designing, shape optimization, pitch-controlling and performance predicting of different wind turbines operated with different control strategies.

For a better understanding of the aerodynamics of the optimized wind turbine rotor model and validation of its performance, a RANS numerical simulations were performed. The overset grid technique helps in predicting the performance of the pitched controlled turbine. With the described simulations approach, the power coefficient is in agreement with the experimental data for the angles of attack below stall. The simulations give a clear indication for the beginning of the stall. The three-dimensional character of the flow on the blade, especially near the root and tip is shown in the simulations. The study showed that the trailing edge separation near the root is suppressed by rotational effects, e. g. Coriolis and centrifugal forces.

After validating the applicability of the developed tools. All facilities for achieving the main objective are ready. Experimental investigations are conducted by exposing the optimized wind turbine model to different turbulence contents in the wind tunnel. The main findings are.

9. Conclusions and Outlook

The higher the turbulence level the higher the power coefficient and hence higher rotational speed for the same incoming wind velocity. Turbulence can influences the performance throughout different means. These are; interactions of turbulence scales with the blade surface boundary layer. This in turn, could delay the stall. Thus, suppressing the boundary layer and preventing it from separation and hence enhancing the aerodynamics characteristics of the blade.

Higher turbulence helps in damping the tip vortices. Thus, reduces the tip losses. In addition, adding winglets to the blade tip will reduce the tip vortex, and then it is possible to isolate major part of them for further individual investigations. The study has shown by adding winglets it is not possible to suppress all tip vortex.

Furthermore, high turbulence content in the incoming wind serves in increasing the wake-surrounding interaction, and hence more energy entrainment to the wake regime. More energetic turbulent flow has been shown to penetrate through the turbine blades, which brings more power in the near-wake regime which cause a faster retrieving of wake.

9.2. Outlook

The present work has shown many observations some were already interpreted and other need intensive investigations.

* High levels of turbulence that can be generated with using of active grid for mimicking the atmospheric one can be performed.

* Dimensionless analysis will help in understanding the scale interaction with the turbine and the turbine diameter, chord and boundary layer scales.

* Applying of different winglets designs in both experimental and numerical for obtaining the optimum design that can extract the maximum power and reduced the vibration and acoustic effects under the influence of turbulence.

* Different Hybrid optimization methods for the sectional airfoil in combination to the rotor shape optimization of TMASO can be performed under the influence of turbulence to harness the maximum possible power from the wind as much as possible.

9. Conclusions and Outlook

* experimental investigations for a wind turbine airfoil in both static and rotating configuration with surface pressure tabs can provide details of boundary layer separation and general pressure distribution.

* Performing numerical simulations with LES can highlight more details about the three dimensionality flow and boundary layer behaviour.

Bibliography

[1] *http://cenvironment.blogspot.de/*

[2] *http://de.slideshare.net/Jupiter276/1-9th-lesson*

[3] *http://user.windsim.com/*

[4] *http://www.intechopen.com/*

[5] *http://www.mathworks.de/*

[6] *http://www.ni.com/*

[7] *http://www.renewableenergyworld.com/*

[8] *http://www.technologyreview.com/energy/*

[9] *http://www.treehugger.com/renewable-energy/*

[10] *http://www.wind-works.org/*

[11] *Turbulent flows*. Cambridge University Press, 2000

[12] *The effect of turbulence intensity on stall of the NACA 0021 aerofoil.* 2001

[13] *Optimization Toolbox User's Guide.* MathWorks, Inc., 2001

[14] *Grundlagen der Strömungsmechanik.* Springer-Verlag, Berlin, Heidelberg, 2006

[15] *Metaheuristics.* Nelson and Henderson (eds.), 2006

[16] *Verification of a new model to calculate turbulence intensity inside a wind farm.* 2006

[17] *Wind blade chord and twist angle optimization using genetic algorithms.* 2006

[18] *Effects of turbulence intensity on power output of wind turbine operating in wake.* 2010

[19] *Usage of Numerical Optimization in Wind Turbine Airfoil Design.* 2010

Bibliography

[20] AL-ABADI, A. ; ERTUNC, Ö. ; WEBER, H. ; DELGADO, A.: A Torque Matched Aerodynamic Performance Analysis Method for the Horizontal Axis Wind Turbines. In: *WIND ENERGY* (2013)

[21] ANJURI, E.R.: Comparison of Experimental results with CFD for NREL Phase VI Rotor with Tip Plate. In: *INTERNATIONAL JOURNAL of RENEWABLE ENERGY RESEARCH* (2012)

[22] ANONYMOUS: Wind Turbine Blade Design Optimization. In: *AIAA* (2010)

[23] BAMPALAS, N. ; GRAHAM, J.M.R.: Aerodynamic Rotor Model for Unsteady flow and Wake Impact / Oldenburg University work shop. June,2008. – Forschungsbericht

[24] BETZ, A.: *Wind-Energie und Ihre Ausnutzung durch Windmühlen*. Göttingen : Vandenhoeck & Ruprecht, 1926

[25] BEYER, F.: *Optimization and Simulation of Flow over Wind Turbines*, Friedrich-Alexander Universitaet Erlangen-Nuernberg, Diplomarbeit, 2012

[26] BOTTASSO, C.L. ; CAMPAGNOLO, F. ; CROCE, A.: In: *Multi-disciplinary constrained optimization of wind turbines, Multibody System Dynamics* 27 (2012), Nr. 1, 21–53. S.

[27] BOURLIS, Dimitris ; CARRIVEAU, DR. R. (Hrsg.): *A Complete Control Scheme for Variable Speed Stall Regulated Wind Turbines, Fundamental and Advanced Topics in Wind Power*. InTech, 2011

[28] CARCANGIU, C.E.: *CFD-RANS Study of Horizontal Axis Wind Turbines*, Universitae degli Studi di Cagliari, Diss., 2008

[29] CARCANGIU, C.E. ; SOERENSEN, J.N. ; CAMBULI, F. ; MANDAS, N.: CFD-RANS analysis of the rotational effects on the boundary layer of wind turbine blades. In: *Journal of Physics* (2007)

[30] CD-ADAPCO: *Star CCM+ 8.02 User Guide*. 2013

[31] CHAMORRO, Leonardo P. ; PORTÉ-AGEL, Fernando: A Wind-Tunnel Investigation of Wind-TurbineWakes: Boundary-Layer Turbulence Effects. In: *Boundary-Layer Meteorol* (2009)

[32] CIMCA 2006: International Conference on Computational Intelligence for Modelling, Control and Automation, Jointly with IAWTIC 2006: International Conference on Intelligent Agents Web Technologies (Veranst.): *Automatic design and optimization of wind turbine blades*. 2007

Bibliography

[33] DE VRIES, Otto: On the theory of the horizontal-axis wind turbine. In: *Annual Review of Fluid Mechanics* 15 (1983), S. 77–96

[34] Department of Aeronautical and Astronautical Engineering; Universitiy of Illinois at Urbana-Champaign; Urbana, Illinois (Veranst.): *Design of a Tapered and Twisted Blade for the NREL Combined Experiment Rotor.* 1999

[35] DURST, F.: *Grundlagen der Stroemungsmechanik.* Springer Berlin, 2006

[36] ELENI, D.C. ; ATHANASIOUS, T.I. ; DIONISSIOS, M.P.: Evaluation of the turbulence models for the simulation of the flow over a National Advisory Committee for Aeronautics (NACA) 0012 airfoil. In: *Journal of Mechanical Engineering Research* (2012)

[37] ERTUNC Özgür: *Experimental and Numerical Investigations of Axisymmetric Turbulence*, Friedrich-Alexander University, Germany,Erlangen, Diss., 2006

[38] EWEA Conference (Veranst.): *Numerical Analysis of a Local Angle of Attack to HAWT Rotor Blade in Unsteady Flow Conditions.* Nov,2004

[39] FRANDSEN S, Thogersen M.: Integrated fatigue loading for wind turbines in wind farms by combining ambient turbulence and wakes. In: *Journal of Wind Energy* (1999)

[40] FUGLSANG, P. ; MADSEN, H. A.: Optimization method for wind turbine rotors. In: *Optimization method for wind turbine rotors, Journal of Wind Engineering and Industrial Aerodynamics* 80 (1999), Nr. 1-2, S. 191–206

[41] GARREL, A. van: Requirements for a wind turbine Aerodynamics simulation Module - Version 1 / ECN-C-01.099. October,2001. – Forschungsbericht

[42] GASCH, R. ; TWELE, J.: *Windkraftanlagen: Grundlagen, Entwurf, Planung und Betrieb.* B. G. Teubner, 2005

[43] GAUNAA, M. ; JOHANSEN, J.: Determination of the Maximum Aerodynamic Efficiency of Wind Turbine Rotors with Winglets. In: *Journal of Physics* (2007)

[44] GHARALI, K. ; JOHNSON, D.A.: Numerical modeling of an S809 airfoil under dynamic stall, erosion and high reduced frequencies. In: *Wind Energy Group; Department of Mechanical and Mechatronics Enigineering; University of Waterloo; Canada N2L 3G1* (2011), april

Bibliography

[45] GIGUERE, P. ; SELIG, M.S. ; TANGLER, J.L.: Blade design trade-offs using low-lift airfoils for stall-regulated HAWTs. In: *Journal of Solar Energy Engineering, Transactions of the ASME* 121 (1999), Nr. 4, S. 217–223

[46] *Kapitel* Windmills and fans. In: GLAUERT, H.: *Aerodynamic Theory*. Bd. 4, Div. L. Springer, Berlin, 1935, S. 324–341

[47] GUNDTOFT, S.: *Wind Turbines*. 2. University College of Aarhus, june 2009

[48] HAND, M.M. ; SIMMS, D.A. ; FINGERSH, L.J. ; JAGER, D.W. ; COTRELL, J.R. ; SCHRECK, S. ; LARWOOD, S.M.: Unsteady Aerodynamics Experiment Phase VI Wind Tunnel Test Configurations and Available Data Campaigns. 2001. – Forschungsbericht. – 8,12 S.

[49] HANSEN, Kurt S. ; BARTHELMIE, Rebecca J. ; JENSEN, Leo E. ; SOMMER, Anders: The impact of turbulence intensity and atmospheric stability on power deficits due to wind turbine wakes at Horns Rev wind farm. In: *Wind Energy* 15 (2012), Nr. 1, 183–196. http://dx.doi.org/10.1002/we.512. – DOI 10.1002/we.512. – ISSN 1099–1824

[50] HIRSCH, C.: *Numerical Computation of Internal and External Flows*. Butterworth-Heinemann, 2008

[51] HUNTER, R. ; FRIIS PEDERSEN, Troels ; DUNBABIN, P. ; ANTONIOU, I. ; FRANDSEN, S. ; KLUG, H. ; ALBERS, A. ; LEE, W.K.: *European wind turbine testing procedure developments. Task 1: Measurement method to verify wind turbine performance characteristics*. 2001 (Riso-R-1209(EN)). – ISBN 87–550–2752–0

[52] HUSKEY, A. ; BOWEN, A. ; JAGER, D.: Wind Turbine Generator System Power Performance Test Report for the Gaia-Wind 11-kW Wind Turbine / NREL/TP-500-46151. 2009. – Forschungsbericht

[53] J. PEINKE, F. Böttcher D. Heinemann B. L. S. Barth B. S. Barth: Turbulence, a challenging problem for wind energy. In: *Journal of Physica A* (2004)

[54] JOHANSEN, J. ; SOERENSEN, N.N.: Aerodynamic investigation of Winglets on Wind Turbine Blades using CFD. In: *Risoe-R Report* (2006)

[55] JONKMAN, J. ; BUTTERFIELD, S. ; MUSIAL, W. ; SCOTT, G.: Definition of a 5-MW Reference Wind Turbine for Offshore System Development. / National Renewable Enuregy Laboratory. 2009. – Forschungsbericht

Bibliography

[56] J.S. DELNERO, F.A. BACCHI J. C. J. MARAON DI LEO L. J. MARAON DI LEO ; BOLDES, U.: EXPERIMENTAL DETERMINATION OF THE INFLUENCE OF TURBULENT SCALE ON THE LIFT AND DRAG COEFFICIENTS OF LOW REYNOLDS NUMBER AIRFOILS. In: *Latin American Applied Research* (2005)

[57] JURECZKO, M. ; PAWLAK, M. ; MEZYK, A.: Optimisation of wind turbine blades. In: *Optimisation of wind turbine blades, Journal of Materials Processing Technology* 167 (2005), Nr. 2-3, S. 463–471.

[58] KANEVČE, G.; Oka S.: Correcting Hot-wire Readings for Influence of Fluid Temperature Variations. In: *DISA Information* (1973)

[59] KIRRKAMM, N. ; STOEVESANDT, B. ; GOLLNICK, B. ; PEINKE, J.: SIMULATION OF A MULTI MEGA WATT WIND TURBINE WITH OPENSOUCE CODE OPEN-FOAM. In: *Poster at European Wind Energy Conference & Exhibition 2010* (2010)

[60] KOOIJMAN, H.J.T. ; LINDENBURG, C. ; WINKELAAR, D. ; HOOFT, E.L. van d.: *Aero-elastic modelling of the DOWEC 6 MW pre-design in PHATAS.* september 2003

[61] LANZAFAME, R. ; MESSINA, M.: Fluid dynamics wind turbine design: Critical analysis, optimization and application of BEM theory. In: *Renewable Energy* 32 (2007), S. 2291–2305

[62] LAURSEN, J. ; ENEVOLDSEN, P. ; HJORT, S.: 3D CFD Quantification of the Performance of a Multi-Megawatt Wind Turbine. In: *Journal of Physics* (2007)

[63] LINDENBERG, C.: *Investigation into Rotor Blade Aerodynamics.* july 2003

[64] LUHUR, M. R. ; WCHTER, M. ; PEINKE, J.: Stochastic modeling of lift and drag dynamics under turbulent conditions. In: *European Wind Energy Conference and Exhibition 2012, EWEC 2012* Bd. 1, 2012, 432-438

[65] MAHU, R. ; POPESCU, F. ; FRUNZULICA, F. ; DUMITRACHE, Al.: 3D CFD Modeling and Simulation of NREL Phase VI ROTOR. In: *Proceedings to American Institute of Physics Conference* (2011)

[66] MANDEZ, G. B. ; ONYEWUDIALA, J. I.: Optimization of Wind Turbine Blades Using Genetic Algorithm. In: *Global Journal of Researches in Engineering (GJRE)* 10 (2010), S. 22

[67] MANWELL, J. F. ; MCGOWAN, J. G. ; ROGERS, A.L.: *Wind Energy Explained: Theory, Design and Application.* John Wiley and Sons, Ltd., 2002

Bibliography

[68] MENTER, F.R. ; KUNTZ, M. ; LANGTRY, R.: Ten Years of Industrial Experience with the SST Turbulence Model. In: *Turbulence, Heat and Mass Transfer* (2003)

[69] MILEY, S.J.: A CATALOGUE OF LOW REYNOLDS NUMBER AIRFOIL DATA FOR WIND TURBINE APPLICATIONS / Department of Aerospace Engineering Texas A&M University. 1982. – Forschungsbericht

[70] MULJADI, E. ; PIERCE, K. ; MIGLIORE, P.: Soft-stall control for variable-speed stall-regulated wind turbines. In: *Soft-stall control for variable-speed stall-regulated wind turbines, Journal of Wind Engineering and Industrial Aerodynamics* 85 (2000), Nr. 3, S. 277–291

[71] National Renewable Eneregy Laboratory (Veranst.): *Wind Turbine Post-Stall Airfoil Performance Characteristics Guidelines for Blade-Element Momentum Methods*. 2004

[72] National Renewable Energy Laboratory (Veranst.): *Rotational Augmentation Disparities in the MEXICO and UAE Phase VI Experiments*. 2010

[73] Proceedings of ASME Turbo Expo (Veranst.): *DEVELOPMENT OF AN EXPERIMENTAL SETUP FOR DOUBLE ROTOR HAWT INVESTIGATION*. Bella Center, Copenhagen, Denmark, June 2012

[74] R. J. BARTHELMIE, M. N. Nielsen S. C. Pryor P.-E. R. S. T. Frandsen F. S. T. Frandsen ; JRGENSEN, H. E.: Modelling and Measurements of Power Losses and Turbulence Intensity in Wind Turbine Wakes at Middelgrunden Offshore Wind Farm. In: *WIND ENERGY* (2007)

[75] ROOIJ, R. van ; TIMMER, N.: *Design of Airfoils for Wind Turbine Blades*. may 2004

[76] S. KANNAN, P. Subbaraj Narayana Prasad P. S. Mary Raja Slochanal S. S. Mary Raja Slochanal: Application of particle swarm optimization technique and its variants to generation expansion planning problem. In: *Electric Power Systems Research* (2004)

[77] S. WATKINS, S. R. ; LOXTON, B.: The Effect of Turbulence on the Aerodynamics of Low Reynolds Number Wings. In: *Engineering Letters* (2010)

[78] SADRAEY, M.: *Aircraft Performance Analysis*. VDM Verlag Dr. Mller, 2009

[79] SANDERSE, B.: Aerodynamics of wind turbine wakes / ECN. 2009. – Forschungsbericht

[80] SCHEPERS, J.G. ; ROOIJ, R.P.J.O.M. van: *Analysis of aerodynamic measurement on a model wind turbine placed in the NASA-Ames tunnel*.

Bibliography

[81] SCHLICHTING, H. ; GERSTEN, K.: *Grenzschicht-Theorie*. Springer Berlin, 2006

[82] The Science of Making Torque from Wind. Jornal of Physics: Conference series (75) (Veranst.): *Design Oriented Aerodynamic Modelling of Wind Turbine Performance*. 2001

[83] SHEN, W.Z. ; SOERENSEN, J.N.: Quasi-3D Navier-Stokes Model for a Rotating Airfoil. In: *Journal of Computational Physics* (1999)

[84] SKENSVED, Erik: *Air Jet for Lift Control in Low Reynolds Number FLow*. 2010

[85] SOMERS, D. M.: Design and Experimental Results for the S809 Airfoil / NREL/SR-440-6918.UC Category:1213.DE97000206. 1997. – Forschungsbericht

[86] STOEVESANDT, B. ; PEINKE, J.: Changes in angle of attack on blades in the turbulent wind field, 2009, 4329-4332. – cited By (since 1996)0

[87] SUMNER, J. ; WATTERS, C.S. ; MASSON, C.: CFD in Wind Energy: The Virtual, Multiscale Wind Tunnel. In: *Energies* (2010)

[88] TANGLER, J.L ; SOMERS, D. M.: NREL Airfoil Families for HAWTs / updated AWEA. 1995. – Forschungsbericht

[89] TÜRK, Matthias ; EMEISB, Stefan: The dependence of offshore turbulence intensity on wind speed. In: *Journal of Wind Engineering and Industrial Aerodynamics* (2010)

[90] VERMEER, L.J. ; SOERENSEN, J.N. ; CRESPO, A.: Wind turbine wake aerodynamics. In: *Progress in Aerospace Sciences* (2003)

[91] VERSTEEG, H.K. ; MALALASEKERA, W.: *An Introduction to Computational Fluid Dynamnics*. Pearson education Limited, 2007

[92] WILSON, R. E. ; LISSAMAN, P. B. S.: Applied aerodynamics of wind power machines / Oreg. St. Univ., Corvallis. 1974 (NSF/RA/N-74113). – Forschungsbericht

[93] WUSSOW, S. ; SITZKI, L. ; HAHM, T.: 3D-simulation of the turbulent wake behind a wind turbine. In: *Journal of Physics* (2007)

[94] Y. KAMADA, J. MURATA T. T. T. MAEDA M. T. MAEDA ; TOBUCHI, A.: Effects of turbulence intensity on dynamic characteristics of wind turbine airfoil. In: *Journal of Science and Technology* (2011)

[95] ZAHLE, F. ; SOERENSEN, N.N. ; JOHANSEN, J.: Wind Turbine Rotor-Tower Interaction Using an Incompressible Overset Grid Method. In: *WIND ENERGY* (2009)

Bibliography

[96] ZAVADIL, R. M.: Wind Generation Technical Characteristics for the NYSERDA Wind Impact Study / EnerNex Corporation. 2003. – Forschungsbericht

Appendix

A.1. Specifications of the NREL UAE phase-VI

The NREL phase-VI wind turbine is a modified version of the previous phases II to V. It consists of two blades tapered and twisted blade wind turbine with the same sectional profile of S809, shown in Figure A.1. This profile is designed for operating under high Reynolds numbers (in a range of 10^6). At low Reynolds numbers the thickness of the S809 leads to a relatively enlarged drag coefficient [44]. Along the rotor, distributions of the chord length c and summation of tip pitch angle and twist angle $\beta = \beta_0 + \beta_t$ are illustrated in Figure A.2. Tip pitch angle is set to different values $\beta_o = 0 - 7°$, the value of $3°$ is a reference for most of the available experimental data. Maximum chord length of $0.737m$ appears at 0.25 blade length and blade twist angle decreases from $21.88°$ at 0.25 blade length to nearly zero degree at the tip [34].

Configuration of blade shape of the NREL UAE phase-VI rotor was considered in the present work. The turbine was optimized to capture the maximum Gross Annual Energy Production (GAEP) with a particular attention for obtaining smooth stall characteristics along the entire blade span [34].

The Rayleigh wind distribution with an average wind velocity of 7.2 m/s was considered in computing GAEP. In contrast to modern control strategies designs, which

A. Appendix

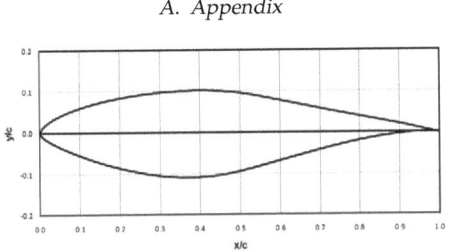

Figure A.1.: S809 Airfoil profile [4].

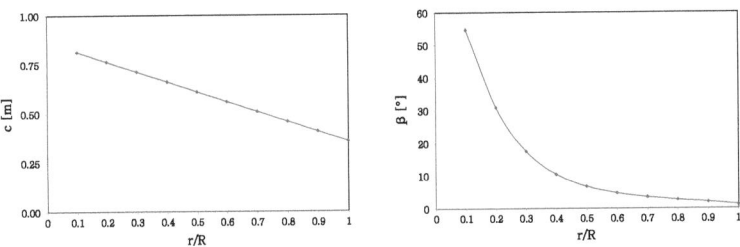

Figure A.2.: Chord and twist distributions over blade length

are pitch-regulated systems, this turbine is stall-regulated and it has a rating power up to 20kW [48]. The rotor has a diameter of $10.058m$ and is composed of two blades. Rotational speed operates constant at $71.6rpm$ [72]. The full turbine is shown in figure A.3.

Figure A.3.: UAE phase-VI turbine mounted in NASA Ames $24.4m$ x $36.6m$ wind tunnel [72].

National Renewable Energy Laboratory analyzed this turbine by instrumenting one

A. Appendix

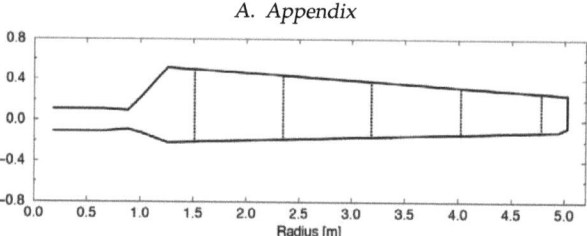

Figure A.4.: UAE phase-VI rotor blade and pressure tabs defined as dotted lines [63].

blade with twenty two chordwise pressure tabs at five spanwise locations, namely at 30%, 47%, 63%, 80%, and 95% span in NASAs 24.4 by 36.6 meter wind tunnel [72]. Geometry of the UAE phase-VI rotor blade and the pressure tabs are specified in Figure A.4.

In Figure A.5 curves of power coefficient over tip speed ratio are shown for different tip pitch angles for the UAE phase-VI rotor. Maximum power coefficient for a tip pitch angle of 3° is reached at $\lambda = 5.2$.

In Figure A.6 experimental torque and power over wind velocity of the UAE phase-VI rotor at a tip pitch angle of 3° are depicted. Rated power P_N is about $10kW$. Rated torque T_N results from dividing rated power by design angular speed ($\omega_o = 2\pi n/60$):

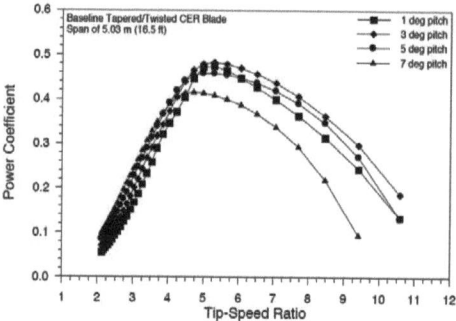

Figure A.5.: Power coefficient over tip speed ratio for the UAE phase-VI rotor at different tip pitch angles [34].

$$T_N = \frac{P_N}{\omega_o} = \frac{10kW}{2\pi\ 71.6\ rpm/60} = 1334\ Nm \qquad (A.1)$$

A. Appendix

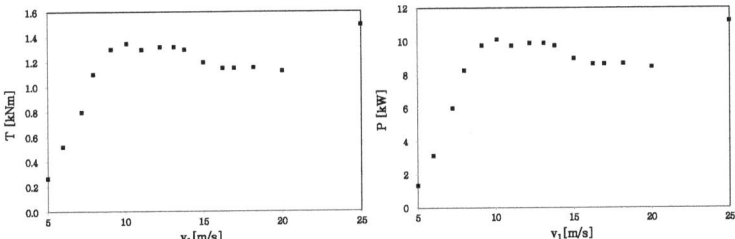

Figure A.6.: Experimental torque and power over wind velocity of the UAE phase-VI rotor at a tip pitch angle of 3° [61]

B
Appendix

B.1. Specifications of NREL 5MW

The offshore installed $5MW$ turbine is shown demonstratively in Figure B.1. Several projects and studies based their calculations on this turbine. For example, the land-based Wind Partnerships for Advanced Component Technology (WindPACT) series of studies, considered wind turbine systems rated up to 5MW. Also the Dutch Offshore Wind Energy Converter (DOWEC) project based its conceptual design on a wind turbine with a 6-MW rating [55].

$5MW$ is the rating power of the turbine, which is achieved at rated rotational speed of $12.1 rpm$ and rated wind velocity of $11.4 m/s$. Cut-in wind velocity is $3m/s$ and it is corresponding to cut-in rotational speed of $6.9 rpm$. Radius of the blade is exactly $61.63m$. This turbine in contrast to the $10kW$ turbine is a pitch-controlled and it is composed of 3 instead of 2 blades. Additionally, instead of one airfoil profile, six different profiles were used along the blade of the $5MW$ turbine. In table B.1 blade aerodynamic properties and spanwise location of the used airfoil profiles are listed. Distribution of chord length and twist angle are shown in Figure B.2 and B.3 respectively. At $15.85m$ maximum chord length of $4.652m$ is reached. From this location chord length decreases nearly linearly to the tip. Blade twist decreases almost linearly from $13.308°$ at the inboard to nearly zero at the tip. In Figure B.4 all shapes of the airfoil profiles are shown,

B. Appendix

which where specified in table B.1. Profiles of $DU35\_A17$ and $DU40\_A17$ are adjusted by leading edge thickness. It is perceptible that the profiles getting narrower from the hub to the tip. Reasons for this design are to increase the glide ratio and to lower the noise at the tip. Blade sections near the tip have a higher insensitivity to roughness than near the hub [75]. More important is to consider the structural demands on the NREL $5MW$ wind turbine because of a twelve times higher blade radius than NREL 10kW, and hence blade weight increment. The airfoil profiles near the hub must be thicker to have the strength to hold the whole blade. In addition to the aforemensioned, airfoil profiles at the tip are thinner to reduce the weight. Therefore, six different airfoil profiles are used along the blade [75].

Figure B.1.: 5MW wind turbine at Hooksiel off the German North Sea coast [7].

In Figure B.5 pitch control angle distribution, generator torque, tip speed ratio and rotational speed over wind velocity of the NREL 5MW wind turbine are depicted. It is obvious that the generator torque and rotational speed are kept constant by the pitch control angle from rated wind velocity at $11.4m/s$.

B. Appendix

Node	RNodes[m]	$\beta[°]$	DRNodes[m]	c[m]	Airfoil profiles
1	2.8667	13.308	2.7333	3.542	$Cylinder1$
2	5.6000	13.308	2.7333	3.854	$Cylinder1$
3	8.3333	13.308	2.7333	4.167	$Cylinder2$
4	11.7500	13.308	4.1000	4.557	$DU40\_A17$
5	15.8500	11.480	4.1000	4.652	$DU35\_A17$
6	19.9500	10.162	4.1000	4.458	$DU35\_A17$
7	24.0500	9.011	4.1000	4.249	$DU30\_A17$
8	28.1500	7.795	4.1000	4.007	$DU25\_A17$
9	32.2500	6.544	4.1000	3.748	$DU25\_A17$
10	36.3500	5.361	4.1000	3.502	$DU21\_A17$
11	40.4500	4.188	4.1000	3.256	$DU21\_A17$
12	44.5500	3.125	4.1000	3.010	$NACA64$
13	48.6500	2.319	4.1000	2.764	$NACA64$
14	52.7500	1.526	4.1000	2.518	$NACA64$
15	56.1667	0.863	2.7333	2.313	$NACA64$
16	58.9000	0.370	2.7333	2.086	$NACA64$
17	61.6333	0.106	2.7333	1.419	$NACA64$

Table B.1.: Distributed blade aerodynamic properties of the NREL $5MW$ wind turbine [55].

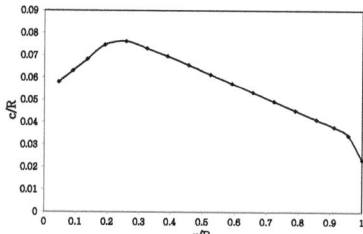

Figure B.2.: Chord distribution of the NREL $5MW$ wind turbine over blade length.

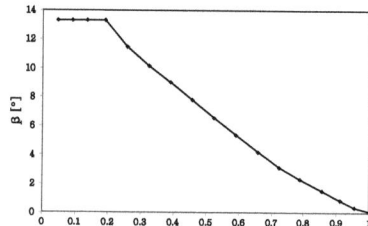

Figure B.3.: Twist distribution of the NREL $5MW$ wind turbine over blade length.

B. Appendix

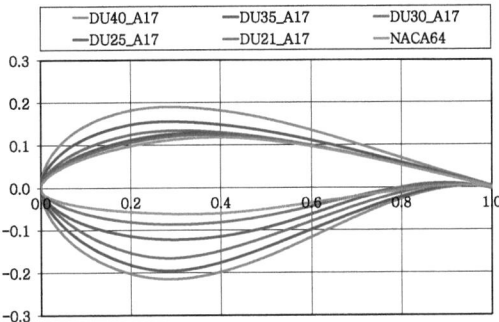

Figure B.4.: Airfoils profiles of the NREL $5MW$ wind turbine from $r = 11.75m$ to the tip [60].

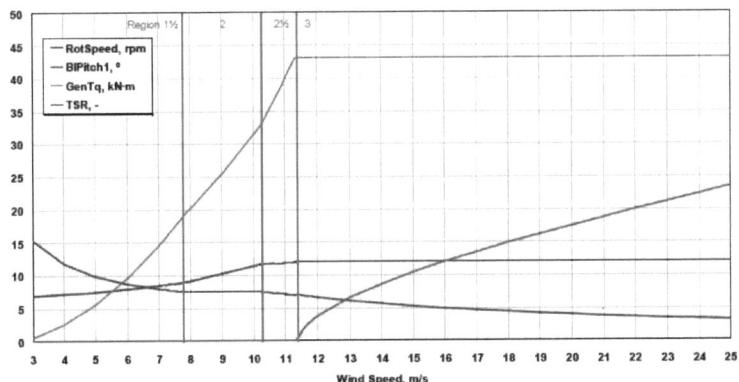

Figure B.5.: Pitch control angle distribution of the NREL 5MW wind turbine [55].

I want morebooks!

Buy your books fast and straightforward online - at one of the world's fastest growing online book stores! Environmentally sound due to Print-on-Demand technologies.

Buy your books online at
www.get-morebooks.com

Kaufen Sie Ihre Bücher schnell und unkompliziert online – auf einer der am schnellsten wachsenden Buchhandelsplattformen weltweit!
Dank Print-On-Demand umwelt- und ressourcenschonend produziert.

Bücher schneller online kaufen
www.morebooks.de

OmniScriptum Marketing DEU GmbH
Heinrich-Böcking-Str. 6-8
D - 66121 Saarbrücken
Telefax: +49 681 93 81 567-9

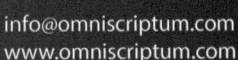
info@omniscriptum.com
www.omniscriptum.com

Printed by Books on Demand GmbH, Norderstedt / Germany